Two Manuals Combined into One Book

Marine Corps Martial Arts

And

Tactical Employment of Nonlethal Weapons

Department of Defense

Pentagon
Publishing

DEPARTMENT OF THE NAVY
Headquarters United States Marine Corps
Washington, D.C. 20380-1775

18 February 1999

FOREWORD

1. PURPOSE

Today's Marines operate within a continuum of force where conflict may change from low intensity to high intensity over a matter of hours. Marines are also engaged in many military operations other than war, such as peacekeeping missions or noncombatant evacuation operations, where deadly force may not be authorized. During noncombative engagements, Marines must determine if a situation warrants applying deadly force. Sometimes Marines must decide in a matter of seconds because their lives or the lives of others depend on their actions. To make the right decision, Marines must understand both the lethal and nonlethal close combat techniques needed to handle the situation responsibly without escalating the violence unnecessarily. Marine Corps Reference Publication (MCRP) 3-02B, *Close Combat*, provides the tactics, techniques, and procedures of Marine Corps close combat. It also provides the doctrinal basis for the Marine Corps Close Combat Training Program (MCCCTP).

2. SCOPE

This publication guides individual Marines, unit leaders, and close combat instructors in the proper tactics, techniques, and procedures for close combat. MCRP 3-02B is not intended to replace supervision by appropriate unit leaders and close combat instruction by qualified instructors. Its role is to ensure standardization and execution of tactics, techniques, and procedures throughout the Marine Corps.

3. SUPERSESSION

MCRP 3-02B supersedes Fleet Marine Force Manual (FMFM) 0-7, *Close Combat*, dated 9 July 1993. There are significant differences between the two publications. MCRP 3-02B should be reviewed in its entirety.

4. WARNING

Techniques described in this publication can cause serious injury or death. Practical application in the training of these techniques will be conducted in strict accordance with approved Entry Level Close Combat, Close Combat Instructor (CCI), and Close Combat Instructor Trainer (CCIT) lesson plans. Where serious danger exists, the reader is alerted by the following:

WARNING

5. CERTIFICATION

Reviewed and approved this date.

BY DIRECTION OF THE COMMANDANT OF THE MARINE CORPS

J. E. RHODES
Lieutenant General, U.S. Marine Corps
Commanding General
Marine Corps Combat Development Command

DISTRIBUTION: 144 000066 00

OVERVIEW OF CLOSE COMBAT

1. Purpose of Close Combat

Close combat is the physical confrontation between two or more opponents. It involves armed and unarmed and lethal and nonlethal fighting techniques that range from enforced compliance to deadly force. The purpose of close combat is to execute armed and unarmed techniques to produce both lethal and nonlethal results. Unarmed techniques include hand-to-hand combat and defense against hand-held weapons. Armed techniques include techniques applied with a rifle, bayonet, knife, baton, or any weapon of opportunity.

2. Continuum of Force

Marines will find themselves in both combative and noncombative situations. The threat level in these situations can rise and fall several times based on the actions of both Marines and the people involved. The escalation of force stops when one person complies with the demands imposed by another person. This range of actions is known as a continuum of force. Continuum of force is the concept that there is a wide range of possible actions, ranging from voice commands to application of deadly force, that may be used to gain and maintain control of a potentially dangerous situation (MCO 5500.6_, *Arming of Security and Law Enforcement [LE] Personnel and the Use of Force*). Continuum of force consists of five levels that correspond to the behavior of the people involved and the actions Marines use to handle the situation (see the table below). Close combat techniques are executed in levels three, four, and five.

Level One: Compliant (Cooperative)

The subject complies with verbal commands. Close combat techniques do not apply.

Level Two: Resistant (Passive)

The subject resists verbal commands but complies immediately to any contact controls. Close combat techniques do not apply.

Continuum of Force		
Level	**Description**	**Actions**
1	Compliant (Cooperative)	Verbal commands
2	Resistant (Passive)	Contact controls
3	Resistant (Active)	Compliance techniques
4	Assaultive (Bodily Harm)	Defensive tactics
5	Assaultive (Serious Bodily Harm/ Death)	Deadly force
Note: Shading indicates levels in which Marines use close combat techniques.		

Level Three: Resistant (Active)

The subject initially demonstrates physical resistance. Marines use compliance techniques to control the situation. Level three incorporates close combat techniques to physically force a subject to comply. Techniques include—

- Come-along holds.
- Soft-handed stunning blows.
- Pain compliance through joint manipulation and the use of pressure points.

Level Four: Assaultive (Bodily Harm)

The subject may physically attack Marines, but he does not use a weapon. Marines use defensive tactics to neutralize the threat. Defensive tactics include the following close combat techniques:

- Blocks.
- Strikes.
- Kicks.
- Enhanced pain compliance procedures.
- Nightstick blocks and blows.

Level Five: Assaultive (Serious Bodily Harm/Death)

The subject usually has a weapon and will either kill or seriously injure someone if he is not stopped immediately and brought under control. Typically, to control the subject, Marines apply deadly force through the use of a firearm, but they may also use armed and unarmed close combat techniques.

3. Marine Corps Tactical Concepts

Close combat techniques support the following key Marine Corps tactical concepts. The concepts are not standalone ideas but are to be combined to achieve an effect that is greater than their separate sum.

Achieving a Decision

Achieving a decision is important in close combat. An indecisive fight wastes energy and possibly Marines' lives. Whether the intent is to control an opponent through restraint or defend themselves in war, Marines must have a clear purpose before engaging in close combat and act decisively once engaged.

Gaining an Advantage

A basic principle of martial arts is to use the opponent's strength and momentum against him to gain more leverage than one's own muscles alone can generate, thereby gaining an advantage. In close combat, Marines must exploit every advantage over an opponent to ensure a successful outcome. This can include employing various weapons and close combat techniques that will present a dilemma to an opponent. Achieving surprise can also greatly increase leverage. Marines try to achieve surprise through deception, stealth, and ambiguity.

Speed

Marines use speed to gain the initiative and advantage over the enemy. In close combat, the speed and violence of the attack against an opponent provides Marines with a distinct advantage. Marines must know and understand the basics of close combat so they can act instinctively with speed to execute close combat techniques.

Adapting

Close combat can be characterized by friction, uncertainty, disorder, and rapid change. Each situation is a unique combination of shifting factors that cannot be controlled with precision or certainty. For example, a crowd control mission may call for Marines to employ various techniques ranging from nonlethal restraint to more forceful applications. Marines who adapt quickly will have a significant advantage.

Exploiting Success

Typically, an enemy will not normally surrender simply because he was placed at a disadvantage. Marines cannot be satisfied with gaining an advantage in a close combat situation. They must exploit any advantage aggressively and ruthlessly until an opportunity arises to completely dominate the opponent. Marines must exploit success by using every advantage that can be gained.

Close Combat

Table of Contents

CHAPTER 1

FUNDAMENTALS OF CLOSE COMBAT

This chapter describes all techniques for a right-handed person. However, all techniques can be executed from either side.

The Marine is depicted in camouflage utilities. The opponent is depicted without camouflage.

The fundamentals of close combat include ranges, weapons of the body, target areas of the body, and pressure points of the body. These fundamentals form the basis for all close combat techniques. They provide Marines with a common framework regardless of the type of confrontation or the techniques used. If Marines apply these fundamentals properly in a close combat situation, they may save their lives or the lives of fellow Marines.

1. Ranges of Close Combat

Close combat engagements occur within three ranges: long range, midrange, and close range. During any engagement, these ranges may blur together or may rapidly transition from one to another until either the opponent is defeated or the conflict is resolved.

Long Range

During long range engagements, combatants engage each other with rifles, bayonets, sticks, or entrenching tools. See figure below.

Midrange

During midrange engagements, combatants engage each other with knives, punches, or kicks.

Close Range

During close range engagements, combatants grab each other. Close range engagements also involve elbow strikes, knee strikes, and grappling.

2. Weapons of the Body

Hands and Arms

The hands, forearms, and elbows are the arm's individual weapons. The hands consist of several areas that can be used as weapons: fists, edges of hands, palms, and fingers.

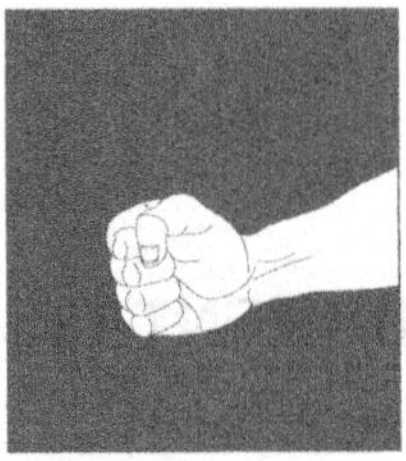

Fists. To minimize injury to the fists, Marines use their fists as weapons to target soft tissue areas such as the throat. The fists' striking surfaces are the first two knuckles of the hands or the meaty portions of the hands below the little fingers.

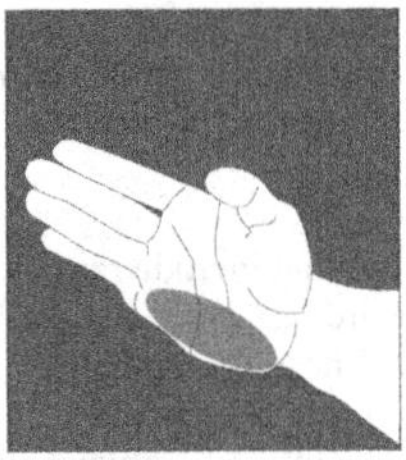

Edge of Hand. Marines use the edge of the hand (knife edge) as a weapon. Marines use the edge of the hand to strike soft tissue areas.

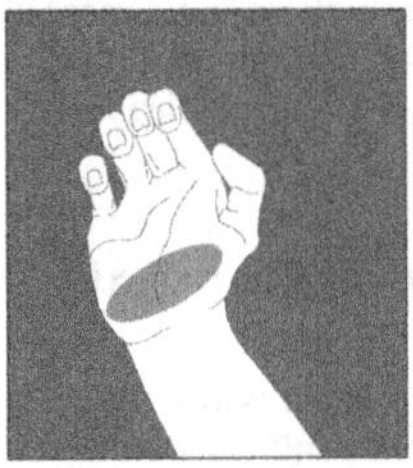

Palms. Because of the palm's padding, Marines use the heels of the palms to strike, parry, and/or block.

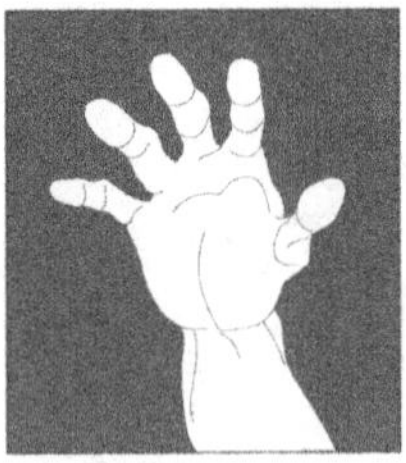

Fingers. Marines use the fingers to gouge, rip, and tear soft tissue areas (e.g., eyes, throat, groin).

Forearms. Marines use the forearms as a defensive tool to deflect or block attacks. Forearms can also be used as striking weapons to damage or break an opponent's joints and limbs. Marines sustain less self-injury when strikes are conducted with the forearms than when strikes are conducted with fists and fingers.

Elbows. Marines use the elbows as striking weapons. Because of the short distance needed to generate power, elbows are excellent weapons for striking during the close range of close combat.

Legs

The legs are more powerful than any other weapon of the body, and they are less prone to injury when striking. The feet are protected by boots and are the preferred choice for striking.

Feet. Marines use the balls of the feet, the insteps, and the toes to kick an opponent. Marines use the cutting edge of the heels and the heels to stomp on an opponent. Marines must be wearing boots when striking with the toes.

Knees. Like elbows, knees are excellent weapons in the close range of close combat. Knee strikes are most effective while fighting close to an opponent where kicks are impractical. The opponent's groin area is an ideal target for the knee strike if he is standing upright. Knee strikes can deliver a devastating secondary attack to an opponent's face following an initial attack that caused him to bend at the waist.

3. Target Areas of the Body

During close combat, Marines strive to attack the accessible target areas of an opponent's body. The readily accessible areas will vary with each situation and throughout the engagement. The target areas are divided into five major groups: head, neck, torso, groin, and extremities. The figure below illustrates target areas of the body.

Head

The vulnerable regions of the head are the eyes, temple, nose, ears, and jaw. Massive damage to the head kills an opponent.

Eyes. The eyes are excellent targets because they are soft tissue areas that are not protected by bone or muscle. Attacks to this area may cause the opponent to protect the area with his hands, allowing Marines to execute a secondary attack to other

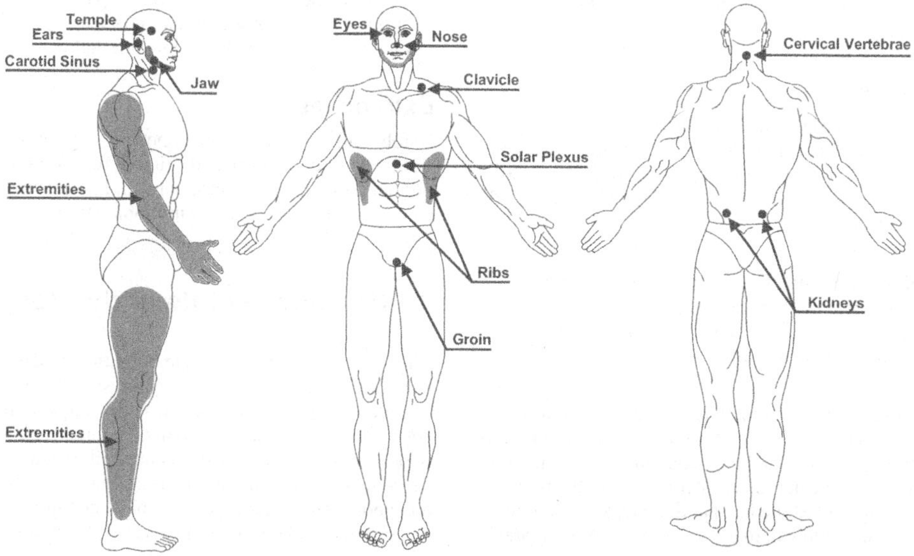

target areas while the opponent uses his hands to protect his eyes.

Temple. The temple is one of the most fragile areas of the head. Powerful strikes to the opponent's temple cause permanent damage and death.

Nose. The nose is very sensitive and easily broken. An attack to this area causes involuntary watering and closing of the opponent's eyes, rendering him vulnerable to secondary attacks. However, through training, individuals can condition themselves to withstand attacks to the nose. Therefore, any attack to the nose must be powerfully delivered.

Ears. Attacks to the ears may cause the eardrum to rupture. But this may not stop or even distract an opponent unless Marines powerfully deliver the strike.

Jaw. The jaw region, when struck forcefully, renders the opponent unconscious. Strikes to the jaw cause painful injuries to the teeth and surrounding tissues (e.g., lips, tongue), but the risk of self-injury is great unless Marines deliver strikes with a hard object such as a helmet, rifle butt, or boot heel.

Neck

The front of the neck, or throat area, is a soft tissue area that is not covered by natural protection. Damage to this region causes the opponent's trachea to swell, closing his airway, which can lead to death.

Carotid Sinus. The carotid sinus is located on both sides of the neck just below the jaw. Strikes to the carotid sinus restrict blood flow to the brain, causing loss of consciousness or death.

Cervical Vertebrae. The cervical vertebrae on the back of the neck, from the base of the skull to the top of the shoulders, contains the spinal cord, which is the nervous system's link to the brain. The weight of the head and the lack of large muscle mass allow damage to the cervical vertebrae and spinal cord. Excessive damage to this area causes pain, paralysis, or death.

Torso

Clavicle. The opponent's clavicle (or collar bone) can be easily fractured, causing immobilization of the arm.

Solar Plexus. Attacks to the opponent's solar plexus or center of the chest can knock the breath out of him and immobilize him.

Ribs. Damage to the opponent's ribs immobilizes him. It may also cause internal trauma.

Kidneys. Powerful attacks to the opponent's kidneys cause immobilization, permanent damage, or death.

Groin

The groin area is another soft tissue area not covered by natural protection. Any damage to this area causes the opponent to involuntarily protect his injured area, usually with his hands or legs. In male opponents, the scrotum is the main target since even a near miss causes severe pain, contraction of the lower abdominal muscles, deterioration of his stance, and possible internal trauma.

Extremities

Rarely will an attack to the opponent's extremities (arms and legs) cause death, but they are important target areas in close combat. Damage to an opponent's joints causes immobilization.

4. Pressure Points of the Body

There are nerves in the human body that, when pressure is applied or when they are struck, allow Marines to control a subject through pain compliance. Marines use pressure points to control an opponent when deadly force is not authorized. They also use pressure points to soften or distract an opponent so a lethal or nonlethal technique can be employed. The figure on page 1-5 illustrates the body's pressure points. Marines execute attacks to pressure points by—

- Rapidly kicking or striking pressure points.
- Slowly applying steady pressure to pressure points.

Infraorbital Nerve

The infraorbital nerve is just below the nose. Marines apply pressure to this nerve with an index finger to control the opponent.

Mastoid Process

The mastoid process is behind the base of the ear and beneath the edge of the jaw. Marines apply inward and upward pressure to this pressure point with the fingers to distract and control the opponent.

Jugular Notch

The jugular notch is at the base of the neck in the notch formed at the center of the clavicle. Marines apply pressure in a quick, stabbing motion with the index finger. Strikes to the jugular notch cause serious damage.

Brachial Plexus (Tie In)

The brachial plexus (tie in) is on the front of the shoulder at the joint. Strikes and pressure applied with the hand are effective on this nerve.

Radial Nerves

Radial nerves are on the inside of the forearms along the radius bones. Strikes and pressure applied with the hand to the radial nerve serve as a softening technique.

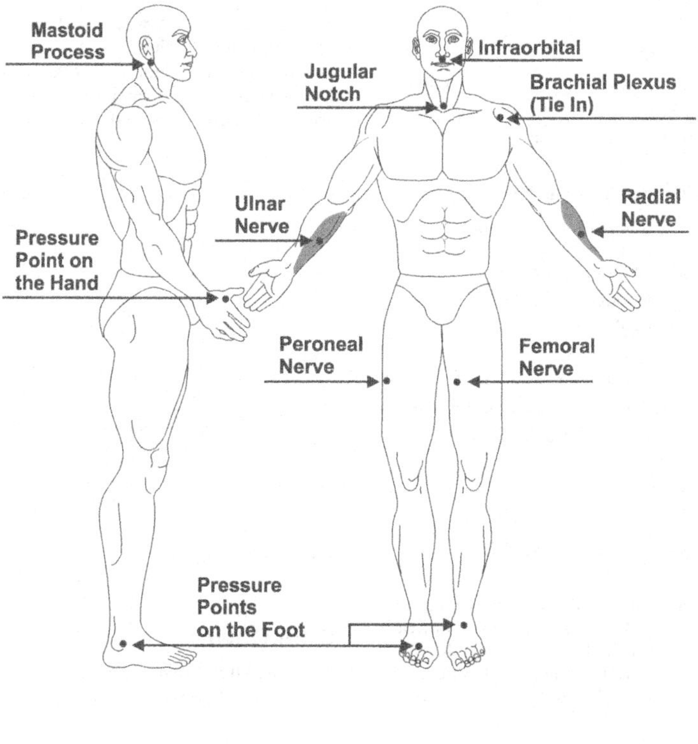

Ulnar Nerve

Ulnar nerves are on the outside of the forearms along the ulnar bones. Strikes and pressure applied with the hand to the ulnar nerve serve as a softening technique.

Pressure Point on the Hand

The hands contain a pressure point on the webbing between the thumbs and index fingers where the two bones of the fingers meet.

To force an opponent to soften or release his grip, Marines apply pressure with their index fingers to this pressure point or strike this pressure point with their fists.

Femoral Nerves

Femoral nerves are on the inside of the thighs along the femur bones. Strikes to the femoral nerve serve as a softening technique.

Peroneal Nerves

Peroneal nerves are on the outside of the thighs along the femur bones. Strikes to the peroneal nerve serve as a softening technique.

Pressure Points on the Feet

There are pressure points on the feet that, when pressure is applied or when they are struck, serve to soften or distract the opponent. Marines apply pressure with the toe, edge, or heel of their boots to the following points:

- The notch below the ball of the ankle.
- The top center of the foot, above the toes.
- The top of the foot where the leg and foot meet.

5. Basic Warrior Stance

Achieving a solid stance is critical to stability and movement throughout any close combat confrontation. The basic warrior stance provides the foundation for stability and movement that is needed to execute close combat techniques. To execute the basic warrior stance, Marines put their feet apart, hands up, elbows in, and chin down.

Feet Apart

Place feet shoulder-width apart.

Keep the head forward and eyes on the opponent, take a half step forward with the left foot, and pivot on the heels so the hips and shoulders are at approximately a 45-degree angle to the right.

Distribute body weight evenly on both legs. Bend the knees slightly.

Hands Up

Curl the fingers naturally into the palm of the hand. Position the thumb across the index and middle fingers. Do not clench the fists. Clenching the fists increases muscular tension in the forearms and decreases speed and reaction time.

Bring the hands up to the face at chin level, with the palms facing each other. Hold the fists up high enough to protect the head, but not so high that they block the field of vision. Ensure continuous eye contact with the opponent.

Elbows In

Tuck the elbows in close to protect the body.

Chin Down

Tuck the chin down to take advantage of the natural protection provided by the shoulders.

6. Angles of Approach and Movement

Marines use movement to control a confrontation and to retain a tactical advantage. Movement increases power and maximizes momentum. By moving around the opponent, Marines gain access to different target areas of the opponent's body. Movement allows Marines to use different weapons of the body and different close combat techniques to attack specific target areas.

Angles of Approach

Marines move anywhere within a 360-degree circle around the opponent to gain a tactical advantage. This circle provides access to different target areas of the opponent's body.

When facing an opponent, Marines move in a 45-degree angle to either side of the opponent. Moving at a 45-degree angle avoids an opponent's strike and puts Marines in the best position to attack the opponent. Marines should avoid being directly in front of an opponent during a confrontation. If a Marine is directly in front of an opponent, the opponent can rely on his forward momentum and linear power to seize the tactical advantage.

Movement

Marines must know how to move in all directions while maintaining the basic warrior stance. During any movement, the legs or feet should not be crossed. Once a movement is completed, the basic warrior stance should be resumed. Maintaining the basic warrior stance protects Marines and puts them in the proper position to launch an attack against an opponent.

Note: Before body movement begins, Marines turn their heads quickly to the new direction. The faster the head turns, the faster the body moves, and the quicker Marines attain visual contact with the opponent.

Forward to the Left. To move forward to the left from the basic warrior stance, Marines—

- Move the left foot forward at a 45-degree angle from the body (approximately 12 to 15 inches), keeping the toe pointed toward the opponent.
- Bring the right foot behind the left foot as soon as the left foot is in place. This returns Marines to the basic warrior stance.

Forward to the Right. To move forward to the right from the basic warrior stance, Marines—

- Move the right foot forward at a 45-degree angle from the body (approximately 12 to 15 inches).
- Bring the left foot, toe pointing toward the opponent, in front of the right foot as soon as the right foot is in place. This returns Marines to the basic warrior stance.

Backward to the Left. To move backward to the left from the basic warrior stance, Marines execute the forward movement in reverse. Marines—

- Move the left foot backward at a 45-degree angle from the body (approximately 12 to 15 inches), keeping the toe pointed toward the opponent.
- Bring the right foot behind the left foot as soon as the left foot is in place. This returns Marines to the basic warrior stance.

Backward to the Right. To move backward to the right from the basic warrior stance, Marines execute the forward movement in reverse. Marines—

- Move the right foot backward at a 45-degree angle from the body (approximately 12 to 15 inches).
- Bring the left foot, toe pointing toward the opponent, in front of the right foot as soon as the right foot is in place. This returns Marines to the basic warrior stance.

7. Balance and Off-Balancing

Balance

In any close combat situation, Marines must strive to maintain balance. The last place to be in a close combat situation is on the ground. Marines must maintain a strong base and a low center of balance, their feet must be a shoulder-width apart, and they must stay on their toes to enable quick movement.

Off-balancing

Marines use off-balancing techniques to control an opponent. These techniques are used to throw an opponent to the ground while Marines remain standing, or they are used to put Marines in a position for an offensive attack.

Off-balancing techniques use the opponent's momentum to move or throw him. For example, if the opponent is charging a Marine, the Marine pulls the opponent to drive him to the ground. Likewise, if the opponent is pulling a Marine, the Marine pushes the opponent to drive him to the ground.

Off-balancing techniques also rely on the power generated by the opponent. For example, during combat a Marine may be tired or outnumbered. Depending on the generated energy and momentum of the opponent, the Marine employs off-balancing techniques with very little effort and still provides effective results.

Because off-balancing techniques rely on the momentum and power generated by the opponent, these techniques are particularly effective for Marines who may be outsized by their opponent or lack their opponent's strength.

Angles of Off-balancing

There are eight angles or directions in which an opponent can be off-balanced: forward, rear, right, left, forward right, forward left, rear right, and rear left.

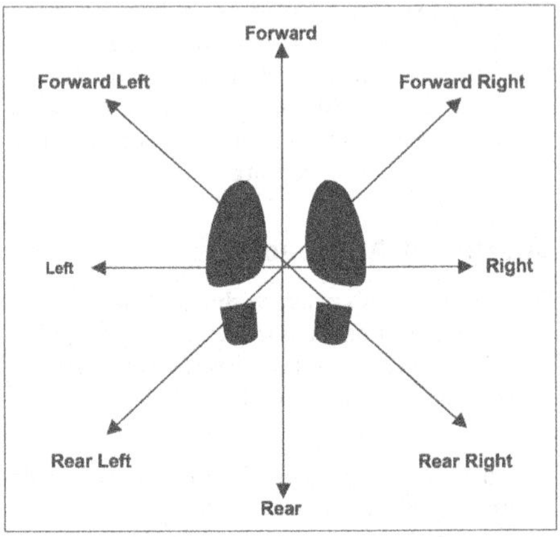

Note: The angles correspond to the Marine's perspective, not the opponent's. Forward, rear, right, and left are straight angles. Forward right, forward left, rear right, and rear left are considered quadrants that are at a 45-degree angle in either direction to the front or rear.

Off-balancing Techniques

Marines off-balance an opponent by pushing, pulling, or bumping the opponent with their hands, arms, or bodies.

To pull, Marines grab an opponent with their hands and drive him forcefully to one of the rear quadrants or to the right or left.

To push, Marines grab the opponent with their hands and drive him forcefully into one of the front quadrants or to the right or left.

Marines execute bumping in the same manner as pushing, but use their shoulders, hips, and legs instead of their hands to off-balance the opponent.

8. Falls

Marines may lose their balance or be thrown to the ground during encounters with an opponent. Marines use falling techniques to absorb the impact of a fall and to quickly return to their feet following an opponent's attack.

Whether falling or being thrown by an opponent, Marines strive to reduce the force of the impact, to prevent serious personal injury, and to increase the chances of survival. Falling techniques use the body's large muscles (back, thighs, buttocks) to protect vital organs and bones from injury and immobilization.

Front Fall

Marines execute a front fall to break a fall on the front. To execute the front fall, Marines—

Bend the elbows and place the palms facing out in a position to spread and absorb the impact of the fall.

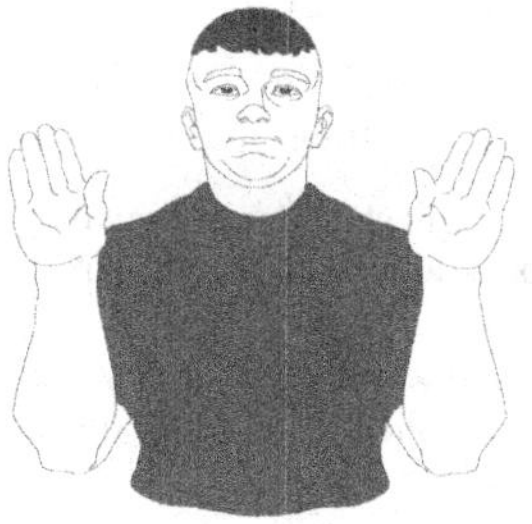

Fall forward, breaking the fall with the forearms and palms. The forearms and hands, down to the fingertips, should strike the ground simultaneously.

Offer resistance with the forearms and hands to keep the head raised off the ground.

Side Fall

Marines execute a side fall to break a fall on the side. To execute the side fall, Marines—

Bring the right arm across the body so the hand is next to the left shoulder with the palm facing inboard.

Fall to the side, breaking the fall with the right arm by slapping the ground and making contact from the shoulder or forearm down to the hand. At the same time, tuck the chin and keep the head raised off the ground. The chin should be tucked to the chest at all times to prevent whiplash.

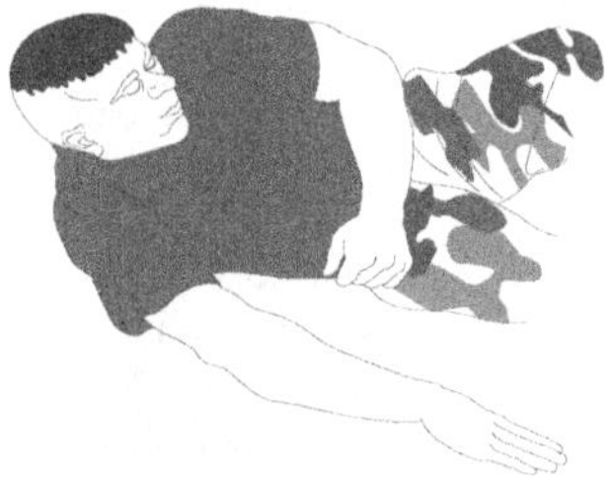

Stretch the right leg out to make contact with the ground and to distribute and absorb the impact. Bend the left leg, allowing the foot to make contact with the ground.

Back Fall

Marines execute a back fall to break the fall when being thrown or falling backward. To execute the back fall, Marines—

Cross the hands in front of the chest and tuck the chin.

Fall backward and slap the ground with the forearms and hands to absorb the impact of the fall and keep the head off the ground.

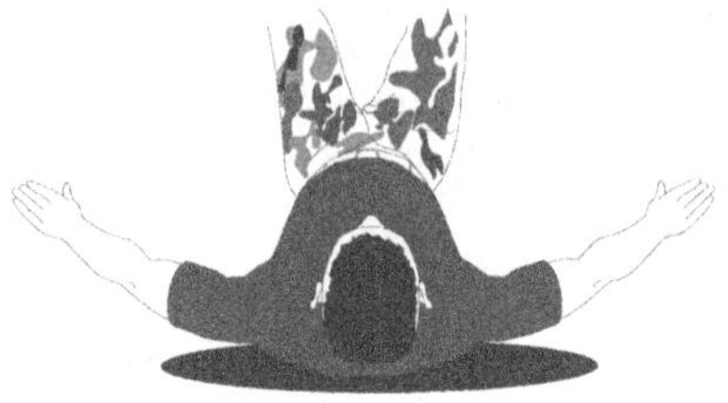

Forward Shoulder Roll

Marines use the forward shoulder roll to break a fall from an opponent's attack and to use the momentum of the fall to get back on their feet quickly. Ideally, Marines execute the forward shoulder roll to a standing position so they can continue fighting. To execute the forward shoulder roll, Marines—

Contact the ground with the back of the right forearm and upper arm. Tuck the chin into the chest.

Roll onto the right shoulder, rolling diagonally across the back to land on the left hip.

Slap the ground with the left arm, absorbing the impact from the shoulder to the hand, palm down.

Keep the left leg straight to absorb as much of the impact as possible. The right leg is bent and the foot hits flat on the ground.

Bend the left leg upon impact to push off with the left knee and leg to a squatting and then a standing position. Forward momentum should carry the Marine to a standing position.

CHAPTER 2

LETHAL AND NONLETHAL WEAPONS TECHNIQUES

This chapter describes all techniques for a right-handed person. However, all techniques can be executed from either side.

In drawings, the Marine is depicted in woodland camouflage utilities; the opponent is depicted without camouflage. In photographs, the Marine is depicted in woodland camouflage utilities; the opponent is depicted in desert camouflage utilities.

1. Bayonet Techniques

WARNING

During training, Marines must have bayonets sheathed. Marines use bayonet dummies to practice bayonet techniques. When practicing offensive and defensive bayonet techniques student-on-student, Marines use pugil sticks.

All Marines armed with a rifle carry a bayonet. The bayonet is an effective weapon if Marines are properly trained in offensive and defensive bayonet techniques. An offensive attack, such as a thrust, is a devastating attack that can quickly end a fight. Defensive techniques, such as the block and parry, can deter the opponent's attack and allow Marines to regain the initiative. Through proper training, Marines develop the courage and confidence required to effectively use a bayonet to protect themselves and destroy the enemy. In situations where friendly and enemy troops are closely mingled and rifle fire and grenades are impractical, the bayonet becomes the weapon of choice.

Holding the Rifle

To execute bayonet techniques, Marines hold the rifle in a modified basic warrior stance. All movement begins and ends with the basic warrior stance. To hold the rifle, Marines—

Use an overhanded grasp to grab the small of the rifle's stock. Use an underhanded grasp to grab the handguards of the rifle.

Lock the buttstock of the rifle against the hip with the right forearm.

Orient the blade end of the rifle toward the opponent.

Offensive Bayonet Techniques

Straight Thrust. Marines use the straight thrust to disable or kill an opponent. It is the most deadly offensive technique because it causes the most trauma to an opponent. Target areas are the opponent's throat, groin, or face. The opponent's chest and stomach are also excellent target areas if not protected by body armor or combat equipment. To execute the straight thrust, Marines—

Lift the left leg and lunge forward off the ball of the right foot while thrusting the blade end of the weapon forward, directly toward the opponent.

Retract the weapon and return to the basic warrior stance.

Slash. Marines use the slash to kill an opponent or to create an opening in his defense. The target area is the opponent's neck. To execute the slash, Marines—

Extend the left hand back toward the left shoulder.

Thrust the left hand forward and swing it to the right, bring the right hand back toward the hip, and turn the cutting edge of the blade toward the opponent's neck. The movement is a slashing motion so the blade cuts across the opponent's neck.

Horizontal Buttstroke. Marines use the horizontal buttstroke to weaken an opponent's defenses, to cause serious injury, or to set him up for a killing blow. Target areas are the opponent's head, neck, and legs. To execute the horizontal buttstroke, Marines—

Step forward with the right foot and drive the right hand forward. Rotate the hips and shoulders into the strike. Move the left hand back toward the left shoulder.

Strike the opponent with the butt of the weapon.

Vertical Buttstroke. Marines use the vertical buttstroke to weaken an opponent's defenses, to cause serious injury, or to set him up for a killing blow. Target areas are the opponent's groin and face. To execute the vertical buttstroke, Marines—

Step forward with the right foot and drive the right hand straight up.

Pull the left hand back over the left shoulder.

Strike the opponent with the butt of the weapon.

Smash. Marines use the smash as a follow-up technique to the horizontal or vertical buttstroke, particularly if they missed the target. The target area is the opponent's head. To execute the smash following a buttstroke, Marines—

Step forward with the right foot and place the blade end of the weapon over the left shoulder and elevate the right elbow above the shoulders.

Drive the arms straight forward, striking the opponent with the butt of the weapon.

Defensive Bayonet Techniques

Parry. Marines use a parry as a defensive technique to redirect or deflect an attack. A parry is a slight redirection of a linear attack by an opponent; e.g., a straight thrust or a smash. To execute the parry, Marines—

Use the bayonet end of the rifle to redirect the barrel or bayonet of the opponent's weapon.

Lock the weapon against the hip with the right forearm.

Rotate to the right or left, moving the bayonet end of the rifle to parry the opponent's attack. Rotation is generated from the hips.

Redirect or guide the opponent's weapon away from the body by exerting pressure against the opponent's weapon.

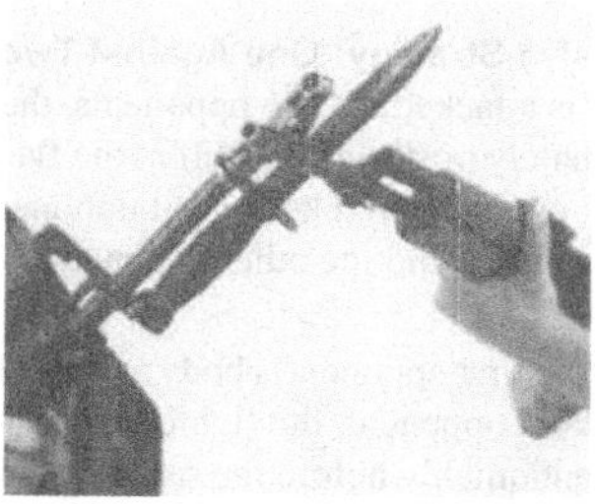

High Block. Marines execute a high block against a vertical attack coming from high to low. To execute the high block, Marines—

Thrust the arms up forcefully at approximately a 45-degree angle from the body. The weapon should be over the top of the head and parallel to the ground. The elbows are bent, but there is enough muscular tension in the arms to absorb the impact and deter the attack.

Low Block. Marines execute the low block

against a vertical attack coming from low to high. To execute the low block, Marines—

Thrust the arms down forcefully at approximately a 45-degree angle from the body. The weapon

should be at or below the waist and parallel to the ground. The elbows are bent, but there is enough muscular tension in the arms to absorb the impact and deter the attack.

Left and Right Block. Marines execute a left or right block against a horizontal buttstroke or a slash. To execute the left or right block, Marines—

Thrust the arms forcefully to the right or left, holding the rifle vertically in the direction of the attack. The elbows are bent, but there is enough muscular tension in the arms to absorb the impact and deter the attack.

Counter Action Following the Block. After deflecting an opponent's attack with a block, Marines counter with a slash or a horizontal buttstroke to regain the initiative. However, the objective in any bayonet fight is to thrust forward with the blade end of the weapon to immediately end the fight.

Group Strategy

On occasion, Marines may engage an opponent as a member of a group or numerous opponents by one's self or as a member of a group. By combining bayonet fighting movements and simple strategies, Marines can effectively overcome their opponent or opponents.

Offensive Strategy: Two Against One. If two bayonet fighters engage one opponent, the fighters advance together.

Fighter 1 engages the opponent while fighter 2 swiftly and aggressively attacks the opponent's exposed flank and destroys the opponent.

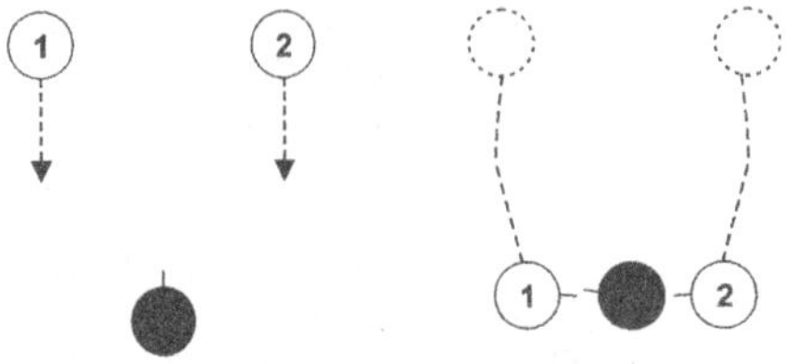

Offensive Strategy: Three Against Two. If three bayonet fighters engage two opponents, the fighters advance together keeping their opponents to the inside.

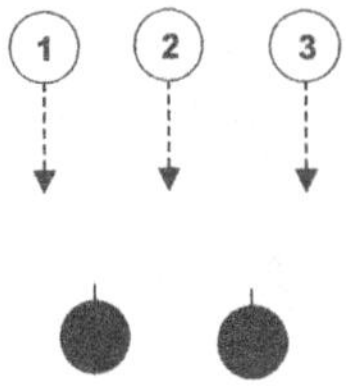

Fighters 1 and 3 engage opponents. Fighter 2 attacks the opponent's exposed flank engaged by fighter 1 and destroys the opponent.

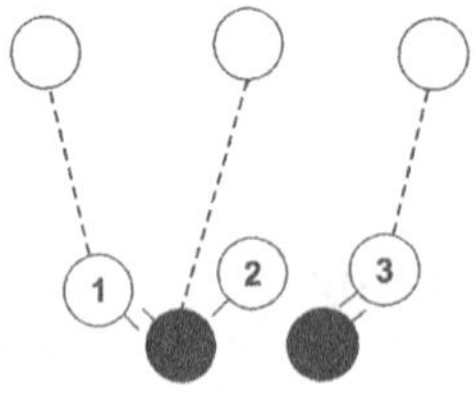

Fighters 1 and 2 turn and attack the exposed flank of the opponent engaged by fighter 3 and destroy the opponent.

Defensive Strategy: One Against Two. If a fighter is attacked by two opponents, the fighter immediately positions himself at the flank of the nearest opponent and keeps that opponent between himself and the other opponent.

Using the first opponent's body as a shield against the second opponent, the fighter destroys the first opponent quickly before the second opponent moves to assist.

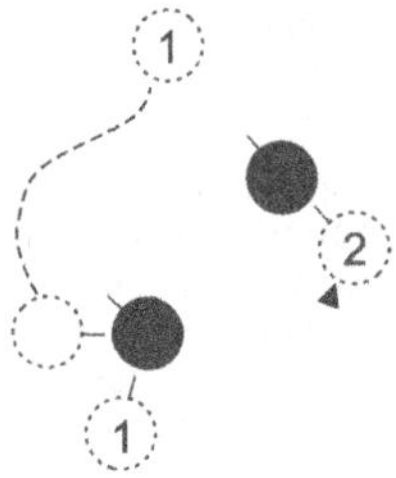

Then, the fighter engages and destroys the second opponent.

Defensive Strategy: Two Against Three. If two fighters are attacked by three opponents, the fighters immediately move to the opponent's flanks.

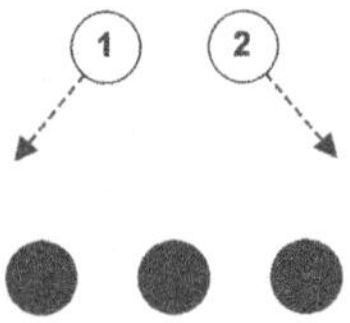

Fighters 1 and 2 quickly attack and destroy their opponents before the third opponent closes in.

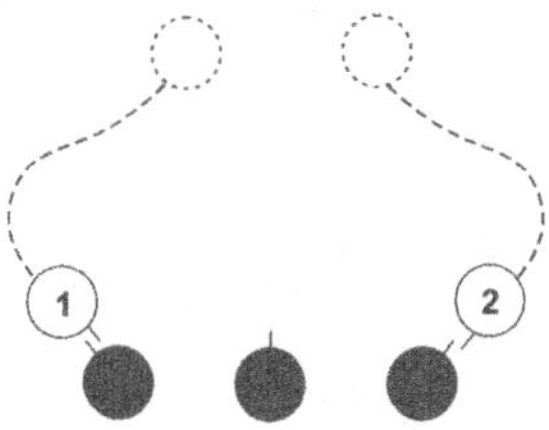

Fighter 1 engages the third opponent while fighter 2 attacks the opponent's exposed flank and destroys the opponent.

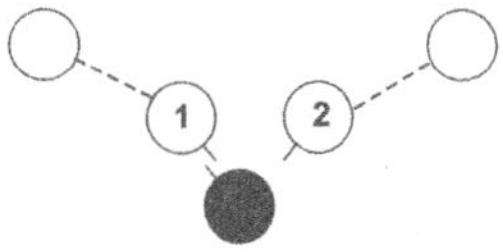

2. Nonlethal Rifle and Shotgun Retention Techniques

Most Marines are armed with the M16A2 service rifle. Marines are taught to keep their weapon with them at all times. Marines must be constantly alert to their surroundings and the people moving in and around their environment. Marines may be confronted by an individual who tries to take their weapons. If this happens, Marines should not struggle with the individual. To retain positive control of their weapons, Marines must understand and apply weapons retention techniques, otherwise known as armed manipulation. The following techniques can be used with either the rifle or the shotgun.

Blocking Technique

To execute a blocking technique, Marines—

Stand in a defensive posture.

Use the weapon to block the opponent by thrusting it out firmly, with the elbows bent. Do not try to hit the opponent with the rifle. The rifle is used as a barrier.

Technique if Opponent Grabs Weapon Underhanded

Usually an opponent will try to grab the weapon or block it as an instinctive action. If the opponent uses an underhand grab to seize the handguards of the weapon, Marines—

Trap the opponent's closest finger(s) above the knuckle with the thumb so he cannot release his grip.

Apply bone pressure on the opponent's finger to initiate pain compliance.

Rotate the barrel of the weapon up or down quickly while maintaining pressure on the opponent's hand. At the same time, quickly pivot to off-balance the opponent.

Technique if Opponent Grabs Weapon Overhanded

If the opponent uses an overhand grab to seize the weapon overhanded, Marines—

Trap the opponent's finger to hold his hand in place.

Rotate the barrel to place it across the opponent's forearm and apply downward pressure. This action is similar to an armbar.

Technique if the Opponent Grabs the Muzzle

If the opponent grabs the muzzle of the weapon, Marines—

Rotate the muzzle quickly in a circle motion. Slash downward with the muzzle to release the opponent's grip.

Butt Strikes

Strikes with the butt of the weapon control or ward off an attacker. During any of the retention techniques, Marines use the heel or cutting edge of the weapon to deliver butt strikes to the inside or the outside of the opponent's thigh. The inside butt strike targets the femoral nerve as illustrated in the photo to the left.

The outside butt strike targets the peroneal nerve as illustrated in the photo to the left. Strikes can be made to the outside or inside of the thighs. If a strike to one side of the thigh misses, Marines follow back through with the butt of the weapon on the other side of the thigh.

Off-balancing Techniques

Marines apply off-balancing techniques to throw an opponent to the ground and retain possession of the weapon.

If the opponent grabs the weapon and pushes, Marines should not push on the weapon. They should—

Move with the momentum and movement of the

opponent by pivoting in the direction of the movement by stepping back.

Throw the opponent to the ground with a quick jerking movement by lowering the muzzle and swinging the butt of the weapon.

If the opponent grabs the weapon and pulls, Marines—

Step on the opponent's foot and push forward to off-balance him and drive him to the ground.

Sweep the opponent's feet out from under him by hooking his leg with the leg and kicking backward.

3. Nonlethal Handgun Retention Techniques

Many Marines are armed with the M9 service pistol. Marines must keep their weapons in their possession at all times. Marines must be constantly alert to their surroundings and the people moving in and around the environment. Marines may be confronted by an individual who tries to take their weapons. To retain positive control of the weapon, Marines must understand and apply handgun retention techniques.

Blocking Technique

Marines perform the following blocking techniques if an opponent attempts to grab their pistol in the holster. Marines—

Place the body between the weapon and the opponent by immediately pivoting so the weapon is away from the opponent.

Step back and away from the opponent while placing the hand on the pistol grip.

Extend the left hand and block, deflect, or strike the opponent's arm.

Armbar Technique

Marines use the armbar technique when an opponent uses his right hand to grab their pistol in the holster. To execute the armbar technique, Marines—

Trap the opponent's right hand by grasping the opponent's wrist or hand with the right hand and applying pressure against the body.

Step back with the right foot and pivot sharply to the right to be next to the opponent. Always pivot in a direction that keeps the weapon away from the opponent.

Straighten the opponent's arm to apply an armbar. The arm should be straight across the torso.

Continue pivoting to the right while pulling back on the opponent's shoulder. This action may break the opponent's arm.

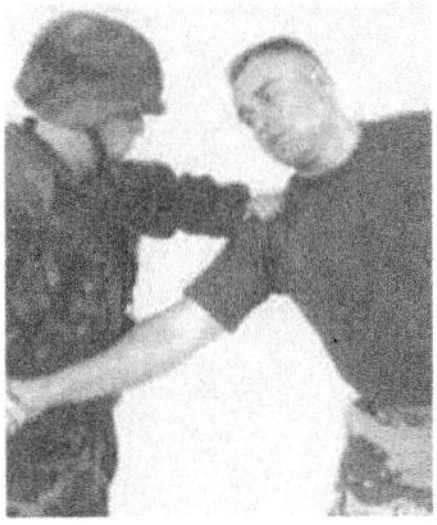

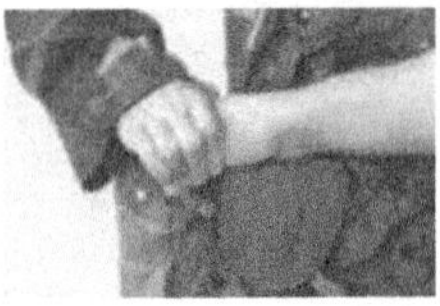

Place the left hand across the opponent's face and apply pressure back and down to take the oppo-

nent to the ground. Pressure applied beneath the nose or on the trachea is equally effective.

Wristlock Technique

Marines use the wristlock technique when an opponent grabs their pistol while it is in the holster with his right hand. To execute the wristlock technique, Marines—

Grasp the opponent's wrist or hand with the right hand and apply pressure against the body.

Step back and away from the opponent and pivot to the right so the weapon is away from the opponent. Always pivot in a direction that keeps the weapon away from the opponent.

Reach over the opponent's arm with the left hand and grab his hand, applying pressure against his arm with the left forearm. Execute a wristlock.

Incorporate the second hand into the wristlock and, stepping back with the left foot, pivot to the left.

Execute a two-handed wristlock by exerting downward pressure with the thumb and rotating his hand to the left.

Continue to pivot to off-balance the opponent and drive him to the ground.

Note: If the opponent grabs the pistol with his left hand, Marines execute the wristlock with one hand and step in toward the opponent, rather than

away from the opponent.

Softening Techniques

Handgun retention techniques use softening techniques applied to pressure points. Bone pressure

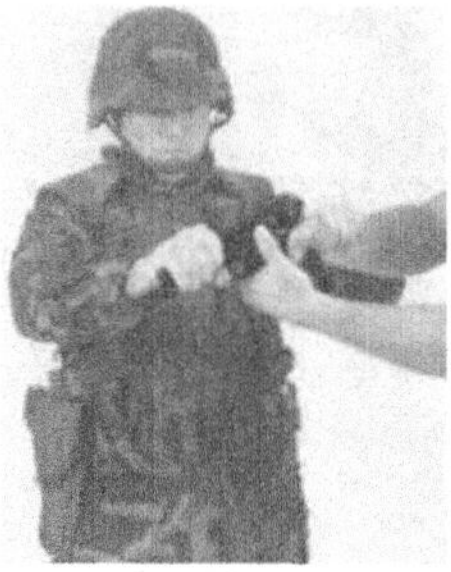

and strikes with the hands (i.e., hammer fist), knees, and feet are also effective softening techniques.

Pressure Points. Marines use pressure point techniques to get the opponent to loosen his grip. Marines use their finger tips to apply pressure to the webbing between the index finger and thumb, the jugular notch, and the brachial plexus tie in. The following figure illustrates pressure applied to the brachial plexus tie in.

Bone Pressure. Bone pressure is the application of pressure on a bone against a hard object to initiate pain compliance. To apply bone pressure, Marines use their hand to trap the opponent's hand on the weapon. Marines apply a slow, steady pressure to the opponent's hand and fingers until his grip is softened or he releases his hold.

Strikes. If it is difficult to apply a retention technique, Marines employ strikes or kicks to force the opponent to loosen his grip. Strikes to the

eyes, the arms (radial nerve), or shoulder (brachial plexus tie in) soften the opponent's grip on the weapon.

Kicks and knee strikes to the peroneal nerve, the femoral nerve, or the groin are effective because the opponent is typically unprepared to counter the strike.

Stomping on the top of the opponent's foot may distract him or loosen his grip on the weapon.

4. Firearm Disarmament Techniques

Marines use firearm disarmament techniques during a close-range confrontation if they are un-

armed and the opponent has a firearm (pistol). These techniques are equally effective if Marines are armed but do not have time to withdraw and present the weapon. The goal of firearm disarmament techniques is to gain control of the situation

so Marines gain the tactical advantage. The goal is not necessarily to get control of the opponent's weapon.

Pistol to the Front

This technique is used when Marines are unarmed and the opponent has a pistol pointing at their front (e.g., head, chest). The technique is the same if the opponent sticks the pistol under the Marine's chin. To execute the counter when an opponent is pointing a pistol toward the front of a Marine, Marines—

Place the hands close to the weapon, about chest high, palms out.

Use the left hand to grab the opponent's forearm and push the opponent's hand with the pistol away to clear the body from in front of the weapon. At the same time, rotate the right shoulder back to clear the body from the weapon.

Maintain control of the opponent's arm.

Grasp the weapon with the right hand by placing the thumb underneath the pistol and the fingers over top of the pistol.

Keep the right hand wrapped tightly around the muzzle and quickly rotate the pistol in the opponent's hand so the muzzle is facing the opponent.

Grasp and pull the opponent's wrist or forearm away from the body while rotating the weapon.

Rotate the weapon toward the opponent while pulling it up and back and out of the opponent's grasp.

Pistol to the Rear

This technique is performed when Marines are unarmed and the opponent has a pistol pointing to the back of the Marine's head. To execute the counter when the opponent is pointing a pistol to the rear, Marines—

Place the hands close together about chest high, palms out.

Step back with the left foot, pivoting on the right foot so the side is against the opponent's front. This action clears the body from the weapon's line of fire. Keep the left hand up.

Pivot on the left foot to face the opponent and, at the same time, raise the left elbow and reach over the top of the opponent's arm with the left arm.

Wrap the left arm tightly around the opponent's arm above his elbow to control it.

Push on the opponent's shoulder with the right hand while pulling up with the left arm to achieve an armbar. This action releases the opponent's grip on the weapon. If necessary, the opponent can be taken to the ground with a leg sweep.

Note: To execute this technique, the weapon must be close to or touching the rear of the Marine. If the weapon is too far away from the body, this technique would be difficult to execute or it would be ineffective.

CHAPTER 3

HAND-HELD WEAPONS

This chapter describes all techniques for a right-handed person. However, all techniques can be executed from either side.

In drawings, the Marine is depicted in woodland camouflage utilities; the opponent is depicted without camouflage. In photographs, the Marine is depicted in woodland camouflage utilities; the opponent is depicted in desert camouflage utilities.

Marines must know how to defend against attacks when an opponent is either unarmed or armed with a held-held weapon. This chapter addresses the combative use of knives, specific weapons of opportunity, and sticks. However, virtually anything can be used as a hand-held weapon.

1. Fundamentals of Knife Fighting

Marines must be trained in knife fighting techniques. Marines experienced in offensive knife techniques can cause enough damage and massive trauma to stop an opponent. When engaged against each other, experienced knife fighters employ various maneuvers and techniques that are specific to knife fighting. Seldom, if ever, will Marines engage an opponent in a classical knife fight.

Note: When armed with a rifle, Marines are issued a bayonet. When armed with a pistol, Marine are issued a combat knife.

Angles of Attack

There are six angles from which an attack with a knife can be launched:

- Vertical strike coming straight down on an opponent.
- Forward diagonal strike coming in at a 45-degree angle to the opponent.
- Reverse diagonal strike coming in at a 45-degree angle to the opponent.
- Forward horizontal strike coming in parallel to the ground.
- Reverse horizontal strike coming in parallel to the ground.
- Forward thrust coming in a straight line to the opponent.

Target Areas of the Body

During any confrontation, the parts of the opponent's body that are exposed or readily accessible will vary. The goal in a knife fight is to attack the body's soft, vital target areas that are readily accessible (e.g, the face, the sides and front of the neck, the lower abdomen [or groin]).

Neck. Carotid arteries, located on either side of the neck, are good target areas because they are not covered by body armor or natural protection.

Lower Abdomen (or Groin). The lower abdomen (or groin region) is a good target area because it is not covered by body armor.

Heart. The heart, if not covered by body armor, is an excellent target which, if struck, can prove fatal in a matter of seconds or minutes.

Secondary Targets. There are secondary target areas that will cause substantial bleeding if an artery is severed. These target areas are not immediately fatal, but can become fatal if left unattended. Attacks to—

- The legs can cause a great deal of trauma and prove fatal. For example, the femoral artery located on the inside of the thigh is a large artery which, if cut, will cause extensive blood loss.
- The brachial artery, located between the biceps and triceps on the inside of the arm, can cause extensive bleeding and damage.
- The arm's radial and ulnar nerves can cause extensive bleeding and damage.

Movement

Marines can move anywhere within a 360-degree circle around the opponent. This allows accessibility to different target areas of the opponent's body. Marines should avoid being directly in front of an opponent because the opponent can rely on his forward momentum to seize the tactical advantage. If Marines face an opponent, movement is made in a 45-degree angle to either side of the opponent. This angle avoids an opponent's strike and places Marines in the best position to attack an opponent.

Wearing the Combat Knife

Marines must wear the combat knife where it is easily accessible and where it can best be retained. It is recommended the combat knife be worn on the weak side hip, blade down, sharp edge facing forward. Marines can place it behind the magazine pouch where it is easily accessible to them, but not easily grabbed by an opponent.

Grip

The grip on the knife should be natural. Marines grasp the knife's grip with the fingers wrapped around the grip naturally as it is pulled out of its sheath. This is commonly known as a hammer grip. The blade end of the knife is always facing the opponent.

Stance

Marines use the basic warrior stance as the foundation for knife techniques. The left hand is a vertical shield that protects either the ribs or the head and neck. The right elbow is bent with the blade pointing forward toward the opponent's head. This position serves as an index point, where all techniques are initiated.

Principles of Knife Fighting

The following are key principles of knife fighting:

- Execute movements with the knife blade within a box, shoulder-width across from the neck down to the waistline. The opponent has a greater chance of blocking an attack if the blade is brought in a wide, sweeping movement to the opponent.
- Close with the opponent, coming straight to the target.
- Move with the knife in straight lines.
- Point the knife's blade tip forward and toward the opponent.
- Apply full body weight and power in each of the knife techniques. Full body weight should be put into the attack in the direction of the blade's movement (slash or thrust).
- Apply constant forward pressure with the body and blade to keep the opponent off-balanced.

2. Knife Fighting Techniques

Slashing Techniques

Marines use slashing techniques to close with an enemy. Slashing techniques distract the opponent or damage the opponent so Marines can close in. Typically, Marines target the opponent's limbs, but any portion of the body that is presented can become a target.

Vertical Slash Technique. The vertical slash follows a vertical line straight down through the target. To execute the vertical slash, Marines—

Thrust the right hand out and bring the weapon straight down on the opponent, continuing to drag the knife down through the opponent's body.

Maintain contact on the opponent's body with the blade of the knife.

Forward Slash Technique. The forward slash follows a straight line in a forehand stroke, across the target areas of either the neck (high diagonal slash) or abdominal region (low horizontal slash). To execute the forward slash, Marines—

Extend the right hand while simultaneously rotating the palm up until the knife blade makes contact with the opponent.

Snap or rotate the wrist through the slashing motion to maximize blade's contact with the opponent.

Drag the knife across the opponent's body, from right to left, in a forehand stroke. The movement ends with the forearm against the body and the knife at the left hip with its blade oriented toward the opponent.

Reverse Slash Technique. The reverse slash is a follow-up technique to a forward attack. It allows Marines both a secondary attack and the ability to resume the basic warrior stance. The reverse slash follows a straight line in a backhand stroke, across the target areas of either the neck (high diagonal slash) or abdominal region (low horizontal slash). To execute the reverse slash, Marines—

Extend the right hand while simultaneously rotating the palm down until the knife blade makes contact with the opponent.

Snap or rotate the wrist through the slashing motion to maximize the blade's contact with the opponent.

Drag the knife across the opponent's body, from left to right, in a backhand stroke. Maintain contact on the opponent's body with the blade of the knife.

Thrusting Techniques

The primary objective of knife fighting is to insert the blade into an opponent to cause extensive damage and trauma. This is done with a thrusting technique. Thrusting techniques are more effective than slashing techniques because of the damage they can inflict. However, Marines use slashing techniques to close with the enemy so that they are closer to the opponent, which allows them to use the thrusting technique.

Vertical Thrust. The thrusting motion follows a vertical line straight up through the target (low into the abdomen region or high into the neck). To execute the vertical thrust, Marines—

Thrust the right hand toward the target, inserting the knife blade straight into the opponent.

Pull the knife out of the opponent.

Forward Thrust. The forward thrust follows a straight line straight into the opponent's neck (high thrust) or abdominal region (low thrust). To execute the forward thrust, Marines—

Thrust the right hand, palm down, toward the target, inserting the knife blade straight into the opponent.

Rotate the palm up once the knife is inserted to twist the blade.

Drop the right elbow and bring the knife to the opposite side of the opponent's body from where it was inserted. At the same time, rotate the hips and shoulders downward to bring body weight to bear on the attack.

Reverse Thrust. The reverse thrust is a follow-up technique to a forward attack. It allows Marines both a secondary attack and the ability to resume the basic warrior stance. The reverse thrust follows a horizontal line straight into the opponent's neck (high thrust) or abdominal region (low thrust). To execute the reverse thrust, Marines—

Bend the right arm, crossing the arm to the left side of the body.

Thrust the right hand, palm up, toward the target, inserting the knife blade straight into the opponent.

Rotate the palm down to twist the blade once the knife is inserted.

Bring the knife to the opposite side of the opponent's body from where it was inserted.

3. Weapons of Opportunity

During an unarmed close combat situation, Marines use their bodies as weapons, but they should be ready and able to use anything around them as a weapon. For example, Marines could throw sand or liquid in an opponent's eyes to temporarily impair his vision or smash the opponent's head with a rock or helmet. Marines must use whatever means are available and do whatever it takes to take control of the situation and to win, or they face the possibility of losing their lives. Some weapons of opportunity are discussed in the following subparagraphs.

Entrenching Tool

An entrenching tool (E-tool) is commonly carried by Marines. It can be an excellent weapon, especially when sharpened. Marines can use the E-tool to block, slash, and thrust at an opponent.

Tent Pole and Pins

Marines can use tent poles and pins to block, strike, or thrust at an opponent.

Web Belt

Marines can stretch a web belt between their hands to block attacks by an opponent.

Battlefield Debris

Marines can use debris on the battlefield (e.g., sticks, glass, a sharp piece of metal) to cut, slash, or stab an opponent. They can also use other types of debris such as shovel or ax handles, boards, metal pipes, or broken rifles to strike an opponent or apply a choke.

Helmet

A helmet can be used to strike an opponent on an unprotected area like the head and face. Grasp the rim of the helmet and thrust the arms forward, striking the opponent with the top of the helmet.

4. Fundamentals of Combative Stick

On the battlefield, Marines must be ready and able to use anything as a weapon. They must learn and be able to use techniques that can be employed with most weapons of opportunity. Among these techniques are combative stick techniques. Combative stick techniques can be used with a stick, a club, a broken rifle, an E-tool, or even a web belt.

Angles of Attack

There are six angles from which an attack with a hand-held weapon can be launched:

- Vertical strike coming straight down on an opponent.
- Forward diagonal strike coming in at a 45-degree angle to the opponent.
- Reverse diagonal strike coming in at a 45-degree angle to the opponent.
- Forward horizontal strike coming in parallel to the ground.
- Reverse horizontal strike coming in parallel to the ground.
- Forward thrust coming in a straight line to the opponent.

Grip

Grasp the stick about 2 inches from its base.

Stance

The basic warrior stance serves as the foundation for combative stick techniques. The left hand becomes a vertical shield that protects the ribs or the head and neck. Depending on how heavy the weapon is, it should be held at a level approximately shoulder height.

Movement

Movement during combative stick techniques is the same as it is for other close combat techniques. Marines can move anywhere within a 360-degree circle around the opponent. This allows accessibility to different target areas of the opponent's body and gains a tactical advantage. Marines should avoid being directly in front of an opponent because the opponent can rely on his forward momentum to seize the tactical advantage. If Marines face an opponent, movement is made in a 45-degree angle to either side of the opponent. This angle avoids an opponent's strike and places Marines in the best position to attack an opponent.

5. Combative Stick Techniques

Strikes

Strikes are intended to inflict as much damage on an opponent as possible. Striking techniques apply to a weapon of opportunity such as a stick, a tent pole, a club, a broken rifle, an E-tool, or a pipe.

Vertical Strike. To execute the vertical strike, Marines—

Bend the right arm, extending the weapon over the back of the right shoulder.

Rotate the forearm straight down off the elbow to bring the weapon down on the opponent.

Rotate the hips and shoulders forcefully toward the opponent.

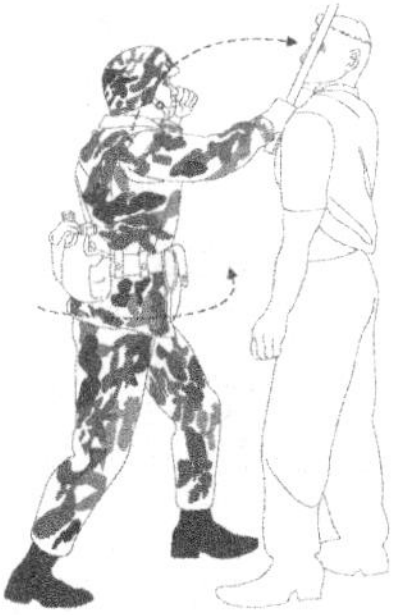

Follow through with the strike by allowing the weight of the weapon to go through the target area of the body.

Forward Strike. To execute a forward strike, Marines—

Step forward with the left foot in the direction of the strike.

Bend the right arm with the elbow extending out to the right and the weapon extended over the right shoulder.

Rotate the forearm to the right of the elbow to bring the weapon down onto the opponent. At the same time, forcefully rotate the hips and shoulders toward the opponent.

Follow through with the strike by allowing the weight of the weapon to go through the target area of the body.

Reverse Strike. The reverse strike is a follow-up technique to a forward strike. It allows Marines both a secondary attack and the ability to resume the basic warrior stance.To execute a reverse strike, Marines—

Step forward with the right foot in the direction of the strike.

Bend the right arm with the hand near the left shoulder. The weapon is extended over the left shoulder.

Rotate the forearm to the right of the elbow to bring the weapon down onto the opponent. At the same time, forcefully rotate the hips and shoulders toward the opponent.

Follow through with the strike by allowing the weight of the weapon to go through the target area of the body.

Forward Thrust. To execute the forward thrust, Marines—

Grasp the stick with the left hand, palm up, in a position where the stick can be controlled with two hands.

Lift the left leg and lunge forward off the ball of the right foot. At the same time, thrust the end of the weapon directly toward the opponent by thrusting both hands forward in a straight line.

6. Blocking Techniques

A block is meant to deter or deflect an attack by an opponent. A block sets up Marines for a follow-on attack against the opponent. Blocks are executed by deflecting, rather than hitting or following through like a strike.

Blocks Against Unarmed Attacks

To block against an unarmed attack, Marines—

Step forward at a 45-degree angle with either the right or left foot. This moves the body out of the line of attack.

Raise the left arm and block the strike with the meaty portion of the forearm.

Employ the stick by one of two techniques:

- Strike the opponent with the stick.
- Use the stick to block on two points of contact. When the stick is used to block, it serves as an incidental strike. This technique is only used if Marines have closed with the opponent and are inside his strike.

Blocks Against Armed Attacks

Block for a Vertical Strike. To execute the block for a vertical strike, Marines—

Step forward with the left foot at a 45-degree angle to the left. This moves the body out of the line of attack.

Block on two points of contact to disperse the impact of the attack:

- Block the opponent's stick by positioning the stick so it is perpendicular to the opponent's stick. If the stick is not perpendicular to the opponent's stick, the stick can slide through and make contact on the Marine.
- Block the opponent's wrist or forearm with the back of the left forearm.

Note: Use the stick to block the opponent's arm if closer to the opponent. It is the same movement as blocking with the arm, except the opponent's arm is blocked with both the stick and the arm.

Block for a Forward Strike. To execute the block for a forward strike, Marines—

Step forward with the left foot at a 45-degree angle to the left. This moves the body out of the line of attack and inside the opponent's strike.

Block on two points of contact to disperse the impact of the attack:

- Block the opponent's wrist or forearm with the meaty portion of the back of the left forearm.
- Strike the opponent's attacking biceps with the stick.

Block for a Reverse Strike. To execute the block for a reverse strike, Marines—

Step forward at a 45-degree angle with the right foot to the right. This moves the body out of the line of attack and inside the opponent's strike.

Block on two points of contact to disperse the impact of the attack:

- Block the opponent's stick by positioning the stick so it is perpendicular to the opponent's stick.

Block the opponent's forearm with the meaty portion of the left forearm.

Note: If closer to the opponent, block his triceps with the back of the left forearm and strike his forearm with the stick.

7. Unarmed Against Hand-Held Weapons

If Marines are engaged against an opponent with a knife, a stick, or some other weapon of opportunity, they must establish and maintain an offensive mindset, not a defensive mindset. Their survival depends on it. They cannot afford to think about getting cut or hurt.

Angles of Attack

Before Marines can learn to block or counter an attack with a hand-held weapon (e.g., knife, stick), they must know from what angle the opponent is attacking. There are six angles from which an opponent will typically attack with a hand-held weapon:

- Vertical strike coming in straight down on the Marine.
- Forward diagonal strike coming in at a 45-degree angle to the Marine.
- Reverse diagonal strike coming in at a 45-degree angle to the Marine.
- Forward horizontal strike coming in parallel to the ground.
- Reverse horizontal strike coming in parallel to the ground.
- Forward thrust coming in a straight line to the Marine.

Blocks

If the opponent has a hand-held weapon (e.g., knife or stick), Marines parry the opponent's hand or arm to block the attack.

Basic Block. To execute the basic block technique, Marines—

Step forward at a 45-degree angle to move out of the line of the attack. Always step in the direction of the strike.

Thrust the forearms forward, hands up, against the opponent's attacking arm. Contact is made on the opponent's arm with the backs of the forearms.

Block for a Vertical Strike. To execute the block against a vertical strike, Marines—

Step forward with the left foot at a 45-degree angle to the left to move out of the line of attack.

Thrust the forearms forward, hands up, against the outside of the opponent's attacking arm.

Block for a Forward Strike. To execute the block against a forward diagonal or forward horizontal strike, Marines—

Step forward with the left foot inside the opponent's attacking arm.

Block the attack with both arms bent so the forearms make contact with the opponent's biceps and forearm.

Block for a Reverse Strike. To execute the block against a reverse diagonal or reverse horizontal strike, Marines—

Step forward with the right foot to the outside of the opponent's attacking arm.

Block the attack with both arms bent so the forearms make contact with the opponent's triceps and forearm.

Block for a Forward Thrust. To execute the block against a forward thrust, Marines—

Bend at the waist, move the hips backward, and jump backward with both feet to move away from the attack. This action is known as "hollowing out."

Hollow out and block the attack with the arms bent and hands together on top of the attacking arm.

Overlap the hands slightly so one thumb is on top of the other hand's index finger. The other thumb should be under the other hand's index finger.

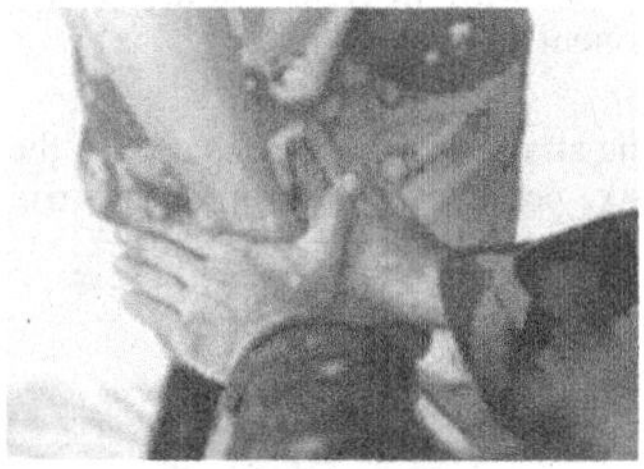

8. Counters to Hand-Held Weapon Attacks

Principles of Counters

A counter is used to control the situation to regain the tactical advantage and end the fight. Regardless of the type of weapon or angle of attack, the following principles apply to countering the attack with a hand-held weapon:

- Move out of the line of attack. Movement is executed in a 45-degree angle forward to the left or right.
- Block the attack.

Note: The first two actions are taken simultaneously.

- Take control of the weapon by controlling the hand or arm that is holding the weapon. Never attempt to grab the opponent's weapon.
- Execute the appropriate follow-up to end the fight; e.g., strikes, joint manipulations, throws, or takedowns (see chap. 8). Marines should continue to attack the opponent until the fight ends.

Counter Techniques

There are two techniques that can be used to counter any armed attack: forward armbar counter and reverse armbar counter. These techniques can be used to counter a vertical attack, a forward diagonal strike, or a forward horizontal strike. With minor variations, the same techniques are used to counter reverse strikes. A third technique, the bent armbar counter, is used to counter a vertical attack.

Forward Armbar Counter. To execute the forward armbar counter to an attack coming from a forward strike, Marines—

Step forward with the left foot inside the opponent's attacking arm.

Block the attack with both arms bent so the forearms make contact with the opponent's biceps and forearm.

Slide the left arm over the opponent's forearm and wrap the arm tightly around his arm, trapping the opponent's attacking arm between the biceps and torso.

Place the right hand on the opponent's shoulder or upper arm to further control his arm and to effect an armbar.

Execute an armbar and continue to exert steady pressure against the arm.

Reverse Armbar Counter. To execute the reverse armbar counter to an attack coming from a forward strike, Marines—

Step forward with the left foot inside the opponent's attacking arm.

Block the attack with both arms bent so the forearms make contact with the opponent's biceps and forearm.

Control the opponent's arm with the left hand and pivot to the right so the back is against the opponent's side. Immediately slide the right arm over the opponent's biceps and wrap the arm tightly around his arm, trapping the opponent's attacking arm between the biceps and torso.

Grasp the opponent's wrist with the left hand and twist his thumb away from the body.

Take the opponent to the ground with an armbar takedown if he does not drop the weapon from the wristlock.

Bent Armbar Counter. This counter is particularly effective against a vertical attack. To execute the bent armbar counter, Marines—

Step forward with the left foot inside the opponent's attacking arm.

Block the attack with both arms bent so the forearms make contact with the opponent's biceps and forearm.

Grasp the opponent's forearm with the left hand. At the same time, slide the right arm underneath the opponent's triceps and grasp the opponent's forearm or wrist with the right hand.

Apply downward pressure with the hands against the opponent's forearm to off-balance the opponent.

CHAPTER 4

STRIKES

This chapter describes all techniques for a right-handed person. However, all techniques can be executed from either side.

In drawings, the Marine is depicted in woodland camouflage utilities; the opponent is depicted without camouflage. In photographs, the Marine is depicted in woodland camouflage utilities; the opponent is depicted in desert camouflage utilities.

Strikes are unarmed individual hitting techniques. Strikes use the hands, elbows, knees, feet, and, in some instances, other parts of the body as personal weapons. Marines must know how to execute strikes effectively. They must also know how to counter strikes from an opponent.

1. Principles of Punches

Muscle Relaxation

Muscle relaxation is crucial when executing punches. The natural tendency in a fight is to tense up, which results in rapid fatigue and decreased power generation. Marines who remain relaxed during a close combat situation generate greater speed, which results in greater generation of power. Relaxing the forearms generates speed and improves reaction time. At the point of impact, Marines clench the fist to cause damage to the opponent and avoid injury to the wrist and hand.

Weight Transfer

Weight transfer is necessary to generate power in a punch. Marines accomplish this by—

- Rotating their hips and shoulders into the attack.
- Moving their body mass straight forward or backward in a straight line.
- Dropping their body weight into an opponent. Body mass can be transferred into an attack from high to low or from low to high.

Rapid Retraction

When Marines deliver a punch, rapid retraction of the fist is important. Once the hand has made contact with the target, Marines quickly return to the basic warrior stance. Rapid retraction—

- Returns the hand and arm to the protection afforded by the basic warrior stance.
- Prevents the opponent from being able to grab the hand or arm.
- Permits the hand and arm to be "chambered" or "re-cocked" in preparation for delivering a subsequent punch.

Telegraphing

Telegraphing a strike occurs when body movements inform the opponent of the intention to launch an attack. Staying relaxed helps to reduce telegraphing.

Often, an untrained fighter telegraphs his intention to attack by drawing his hand back in view of his opponent, changing facial expression, tensing neck muscles, or twitching. These movements, however small, immediately indicate an attack is about to be delivered. If the opponent is a trained fighter, he may be able to evade or counter the attack. If the opponent is an untrained fighter, he may still be able to minimize the effect of an attack.

2. Punches

Punches may be thrown during any hand-to-hand confrontation. Most people resort to punching because it is a natural reaction to a threat. The purpose of a punch is to stun the opponent or to set him up for a follow-up finishing technique. However, punches should only be executed to the soft tissue areas of an opponent. A correctly delivered punch maximizes the damage to an opponent while minimizing the risk of injury to Marines.

Basic Fist

Punches are executed using the basic fist. To make the basic fist, the fingers are curled naturally into the palm of the hand and the thumb is placed across the index and middle fingers. Do not clench the fist until movement has begun. This reduces muscular tension in the forearm and increases speed and reaction time. Just before impact, Marines exert muscular tension on the hand and forearm to maximize damage to the opponent and reduce their chances of injury. Contact should be made with the knuckles of the index and middle fingers.

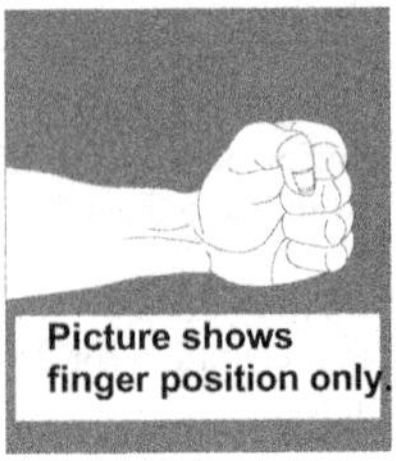

When striking with the basic fist, Marines must keep the hand straight, or in line, with the wrist to avoid injury to the wrist.

Lead Hand Punch

The lead hand punch is a snapping straight punch executed by the forward or lead hand. It is a fast punch designed to keep the opponent away and to set up an attack. A lead hand punch conceals movement and allows Marines to get close to the opponent. Lead hand punches should strike soft tissue areas, if possible. To execute the lead hand punch, Marines—

Snap the lead hand out to nearly full extension, while rotating the palm to the ground.

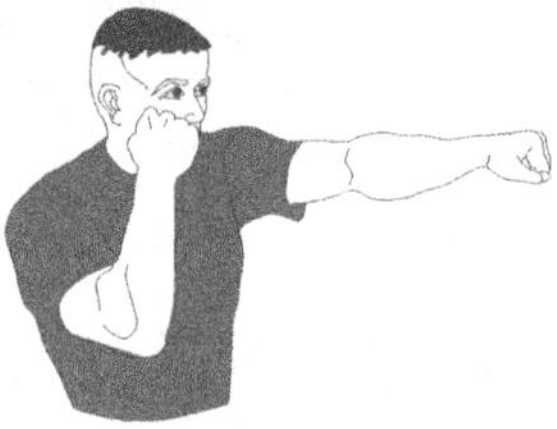

Contact the opponent with the first two knuckles of the fist.

Retract the hand immediately, resuming the basic warrior stance.

Rear Hand Punch

The rear hand punch is a snapping punch executed by the rear (right) hand. It is a power punch designed to inflict maximum damage on the opponent. Its power comes from pushing off the rear leg and rotating the hips and shoulders. To execute the rear hand punch, Marines—

Rotate the hips and shoulders forcefully toward the opponent and thrust the rear hand straight out, rotating the palm down, to nearly full extension.

Shift body weight to the lead foot while pushing off on the ball of the rear foot.

Contact the opponent with the first two knuckles of the fist.

Retract the hand immediately.

Uppercut

The uppercut is a powerful punch originating below the opponent's line of vision. It is executed in an upward motion traveling up the centerline of the opponent's body. It is delivered in close and usually follows a preparatory strike that leaves the target area unprotected. When delivered to the chin or jaw, the uppercut can render an opponent unconscious, cause extensive damage to the neck, or sever the tongue. To execute the uppercut, Marines—

Bend the arms, rotating the palm inboard. The distance the arms bend depends on how close the opponent is.

Rotate the hips and shoulders forcefully toward the opponent, thrusting the fist straight up toward the opponent's chin or jaw.

Contact the opponent with the first two knuckles of the fist.

Retract the hand immediately.

Hook

The hook is a powerful punch that is executed close in and is usually preceded by a preparatory strike. To execute the hook, Marines—

Thrust the right arm in a hooking motion toward the opponent, keeping the elbow bent while forcefully rotating the right shoulder and hip toward the opponent.

Contact the opponent with the first two knuckles of the fist. Continue rotating the shoulder and hip, following through with the fist to the target.

Retract the hand immediately.

3. Strikes with the Upper Body

Strikes stun the opponent or set him up for a follow-up finishing technique. The hands, forearms, and elbows are individual weapons of the arms that can be used to execute strikes including the hammer fist, chin jab, knife hand, eye gouge, and elbow strikes. These strikes provide a variety of techniques that can be used in any type of close combat situation.

Principles of Execution

There are several principles of execution that ensure a strike's effectiveness.

Generating Power. Weight transfer is necessary to generate maximum power in a punch. Marines accomplish this by—

- Rotating the hips and shoulders into the attack.
- Moving the body mass straight forward or backward in a straight line.
- Dropping body weight into an opponent. Body mass can be transferred into an attack from high to low or from low to high.

Muscular Tension. The arms are relaxed until the moment of impact. At the point of impact, Marines apply muscular tension in the hand and forearm to maximize damage to the opponent and to avoid injury to the hand. The arms are relaxed until the moment of impact.

Hit and Stick and Follow-Through. A strike should be delivered so that the weapon (e.g., hand, elbow) hits and remains on the impact site (target) and follows through the target. This technique inflicts maximum damage on the opponent.

Strikes with the arms are executed with "heavy hands"; i.e., the strike is executed by driving through with the strike to allow the weight of the hand to go through the target area of the body. Contact with the opponent should be made with the arm slightly bent, and the arm extends as it moves through the target. This technique allows Marines to deliver strikes effectively without executing full force.

Movement. Movement puts Marines in the proper position for launching an attack against an opponent as well as providing protection.

Movement is initiated from the basic warrior stance and ends by resuming the basic warrior stance. Strikes can be performed with either the left or right arm depending upon—

- The angle of attack.
- The position of the opponent.
- The opponent's available, vulnerable target areas.

Target Areas of the Body

For each strike, there are target areas of the body that, when struck, maximize damage to an opponent. Strikes use gross motor skills as opposed to fine motor skills. The target areas of the body are just that—areas. Pinpoint accuracy on a specific nerve is not needed for the strike to be effective.

Hammer Fist

Striking with the hammer fist concentrates power in a small part of the hand, which, when transferred to the target, can have a devastating effect. The striking surface of the hammer fist is the meaty portion of the hand below the little finger. To execute the hammer fist strike, Marines—

Make a fist and bend the arm at approximately a 45- to 90-degree angle. At the same time, rotate the right hip and right shoulder backward.

Thrust the fist forward onto the opponent while rotating the right hip and shoulder forward.

Rotate the wrist so the hammer fist makes contact on the opponent.

Chin Jab

The chin jab can immediately render an opponent unconscious and cause extensive damage to the neck and spine. The striking surface is the heel of the palm of the hand. To execute the chin jab, Marines—

Bend the right wrist back at a 90-degree angle with the palm facing the opponent and the fingers pointing up.

Keep the right arm bent and close to the body. Extend the hand into a concave position with the fingers slightly spread apart.

Step forward with the left foot toward the opponent, keeping the feet approximately shoulder-width apart and the knees bent. This is done to close with the opponent.

Keep the right arm bent and close to the side. Thrust the palm of the hand directly up under the opponent's chin. At the same time, rotate the right hip forward to drive the body weight into the attack to increase the power of the strike. The attack should travel up the centerline of the opponent's chest to his chin.

Knife Hand

The knife hand is one of the most versatile and devastating strikes. The striking surface is the cutting edge of the hand, which is the meaty portion of the hand below the little finger extending to the top of the wrist. The striking surface is narrow, allowing strikes on the neck between the opponent's body armor and helmet. The knife hand strike is executed from one of three angles: outside, inside, and vertical.

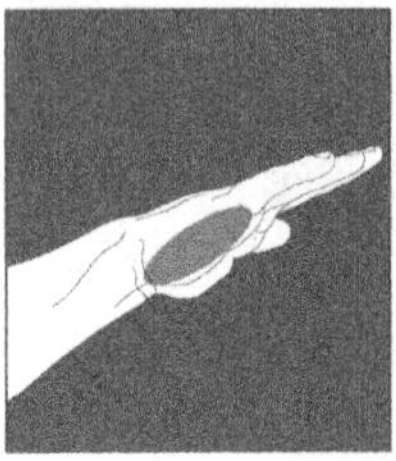

Outside Knife Hand. To execute the outside knife hand strike, Marines—

Execute a knife hand by extending and joining the fingers of the right hand and placing the thumb next to the forefinger (like saluting).

Retract the right hand. At the same time, rotate the right hip and right shoulder backward.

Thrust the knife hand forward (horizontally) onto the opponent while rotating the right hip and shoulder forward.

Inside Knife Hand. To execute the inside knife hand strike, Marines—

Execute a knife hand.

Bring the right hand over the left shoulder. At the same time, rotate the right shoulder forward and the left hip forward.

Thrust the knife hand forward (horizontally) onto the opponent while rotating the right hip and shoulder forward and the left shoulder backward.

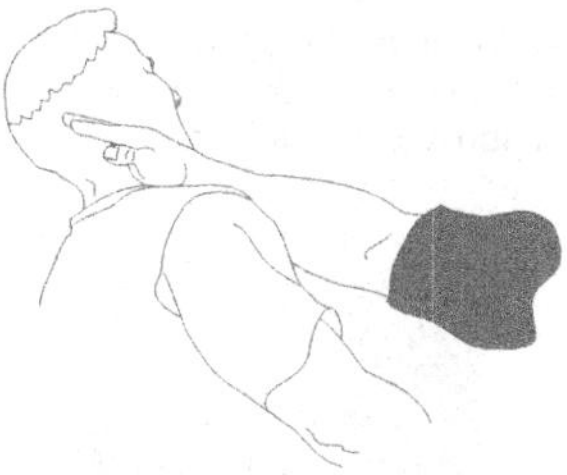

Vertical Knife Hand. When thrown vertically, the knife hand strike comes straight down in a straight line.

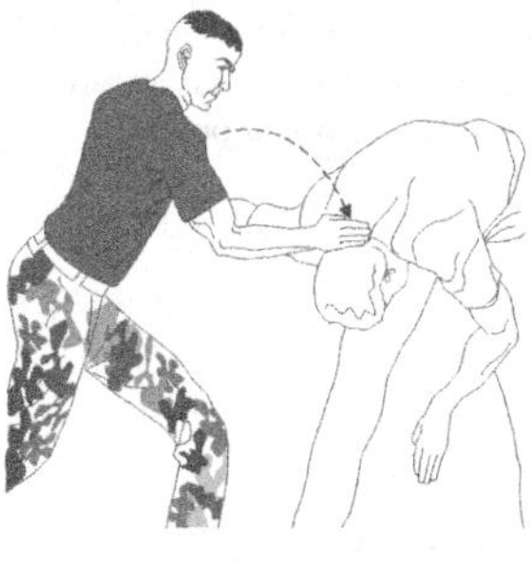

Eye Gouge

The eye gouge is used to attack an opponent's eyes, blinding him so follow-up strikes can be executed. The striking surface is the tips of the fingers or thumb. To execute the eye gouge, Marines—

Extend the right hand with the fingers slightly spread apart to allow entry into the eye sockets.

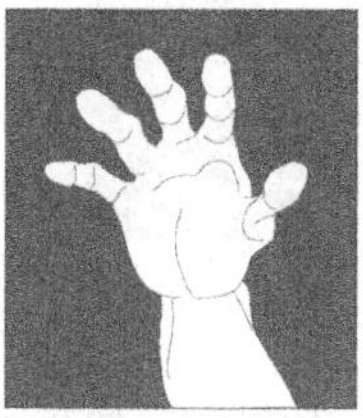

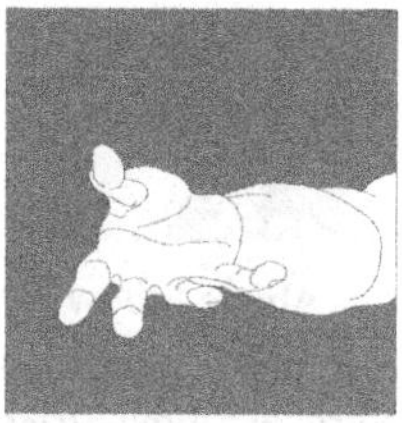

Place the palm of the hand either toward the ground or toward the sky and thrust the right hand forward into the opponent's eyes.

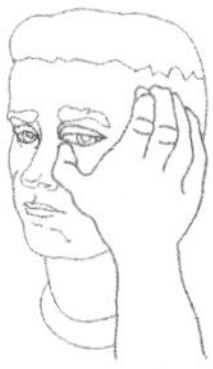

Thrust the hand forward at the opponent's nose level so the fingers or thumb slide naturally into the grooves of the opponent's eye sockets.

Elbow Strikes

The elbow is a powerful weapon that can be used in several different ways to attack virtually any part of an opponent's body. Elbow strikes can be performed either vertically (upward or downward) or horizontally (forward or reverse). The striking surface is 2 inches above or below the point of the elbow, depending upon the angle of attack, the opponent's attack angle, and the position of the opponent.

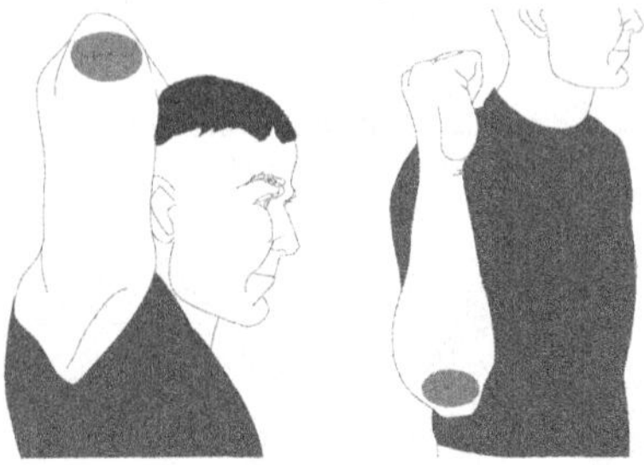

Above Elbow **Below Elbow**

Vertical Elbow Strike (Up). To execute an upward vertical elbow strike, Marines—

Bend the right elbow, keeping the fist close to the body. The fist is at shoulder level and the elbow is next to the torso.

Thrust the elbow vertically upward toward the opponent while rotating the right shoulder and hip forward to generate additional power.

Contact the opponent with the right forearm 2 inches above the point of the elbow.

Vertical Elbow Strike (Down). To execute a downward vertical elbow strike, Marines—

Bend the right elbow, keeping the fist close to the body. The fist is on the shoulder and the elbow is raised well above the shoulder.

Thrust the elbow vertically downward toward the opponent while dropping body weight into the attack to generate additional power.

Contact the opponent with the right triceps 2 inches above the point of the elbow.

Horizontal Elbow Strike (Forward). To execute the forward horizontal elbow strike, Marines—

Tuck the right fist near the chest with the palm heel facing the ground.

Thrust the right elbow horizontally forward toward the opponent. The forearm is parallel to the ground.

Rotate the right shoulder and hip forward to generate additional power.

Contact the opponent with the right forearm 2 inches below the point of the elbow.

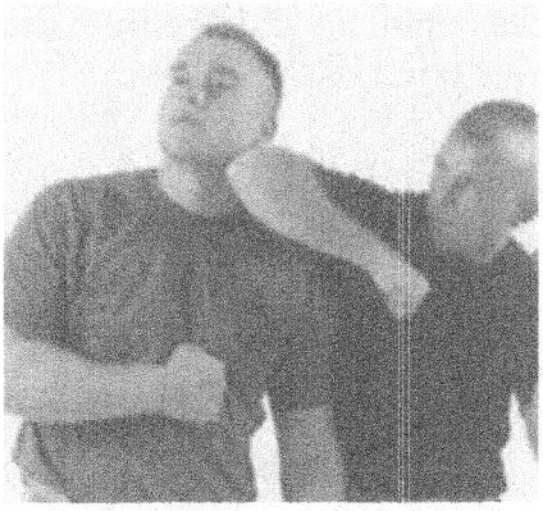

Horizontal Elbow Strike (Rear). To execute the rear horizontal elbow strike, Marines—

Tuck the right fist near the left shoulder with the palm heel facing the ground. At the same time, rotate the right shoulder forward and the left hip forward.

Thrust the right elbow horizontally rearward toward the opponent. The forearm is parallel to the ground and the hand moves toward the direction of the attack.

Rotate the right hip back and the right shoulder backward to generate additional power.

Contact the opponent with the right triceps 2 inches above the point of the elbow.

4. Strikes with the Lower Body

The legs are the body's most powerful weapons because they use the largest muscle groups to generate a strike. Legs are also less prone to injury. The feet are the preferred choice for striking because they are protected by boots. Marines use their feet, heels, and knees to execute kicks, knee strikes, and stomps.

Kicks

The purpose of a kick is to stop an opponent's attack or to create an opening in his defense in order to launch an attack. Kicks can be performed with the left (lead) leg or the right (rear) leg. Kicks with the rear leg have greater power because the hips are rotated into the attack. However, the rear leg is further away from the opponent so a strike with the rear leg will not contact the opponent as quickly as a strike with the lead leg.

Front Kick. The front kick is executed when the opponent is in front of the Marine. The front kick, delivered with either the rear or lead leg, is effective for striking below the waist. Attempting to kick higher results in diminished balance and provides the enemy with a greater opportunity to grab the leg or foot. The striking surfaces are the toe of the boot or the bootlaces. To execute the front kick, Marines—

Raise the left knee waist high, pivot the hips into the attack, and thrust the left foot forward toward the opponent.

Contact the opponent with the toe of the left boot or bootlaces.

Return to the basic warrior stance.

Side Kick. The side kick, delivered with the lead leg, is effective for striking the knees. The side kick is executed when the opponent is to the side of the Marine. The striking surface is the outside cutting edge of the boot near the heel. To execute the side kick, Marines—

Raise the right knee waist high and rotate the right hip forward.

Thrust the right foot to the right side toward the opponent, turning the foot at a 90-degree angle to maximize the striking surface on the opponent.

Contact the opponent with the cutting edge of the right boot.

Return to the basic warrior stance.

Knee Strikes

Knee strikes are excellent weapons during the close range of close combat fighting. The knee strike is generally delivered in close.

Vertical Knee Strike. The striking surface is the thigh, slightly above the knee. To execute the vertical knee strike, Marines—

Raise the right knee and drive it up forcefully into the opponent. Power is generated by thrusting the leg upward.

Contact the opponent 2 inches above the right knee.

Horizontal Knee Strike. The horizontal knee strike is executed with the leg generally parallel with the ground while rotating the hips to generate power. It is often delivered to the peroneal nerve. The striking surface is the front of the leg, slightly above or below the knee. To execute the horizontal knee strike, Marines—

Raise the right knee, rotate the right hip forward while pivoting on the left foot, and drive the knee horizontally into the opponent.

Contact the opponent 2 inches above the right knee.

Stomps

Stomps are delivered with the feet, usually when the opponent is down. Remember, when the opponent is down, Marines take whatever target is available.

Vertical Stomp. The vertical stomp allows Marines to remain upright and balanced, to rapidly deliver multiple blows with either foot, and to quickly and accurately attack the target. It is the preferred stomp. The striking surface is the flat bottom of the boot or the cutting edge of the heel. To execute the vertical stomp, Marines—

Raise the right knee above the waist with the right leg bent at approximately a 90-degree angle.

Drive the flat bottom of the right boot or the cutting edge of the right heel down onto the opponent forcefully. At the same time, bend the left knee slightly to drop the body weight into the strike.

Ax Stomp. The striking surface of the ax stomp is the cutting edge of the heel. To execute the ax stomp, Marines—

Raise the right heel above the waist, keeping the right leg straight.

Drive the cutting edge of the right heel down onto the opponent forcefully. At the same time, bend the left knee slightly to drop the body weight into the strike.

5. Counters to Strikes

In a close combat situation, an opponent will attempt to strike Marines with punches and kicks. When an opponent uses a strike, Marines must first avoid the strike. This is accomplished by moving quickly and blocking. Next, Marines must get into an offensive position. This allows Marines to use offensive strikes to attack the opponent. Regardless of the strike, the counter to a strike requires Marines to move, block, and strike.

Move

The first step in countering a strike is to move out of the way of the strike's impact. Movement removes Marines from the opponent's intended strike point and positions Marines to attack. Movement is executed at approximately a 45-degree angle to the front or rear. Movement is always initiated from the basic warrior stance. Return to the basic warrior stance with the toe of the lead foot pointing toward the opponent once the movement is complete.

Block

Different blocks are executed based on the strike. These will be covered by individual counters.

Strike

Any of the upper body or lower body strikes can be executed as a follow-on attack or part of the counter to an opponent's strike. The follow-on strike is determined by the angle to the opponent, the position of the opponent, and the opponent's available vulnerable target areas.

Counters to Punches

Counter to a Lead Hand Punch. This counter is used when the opponent throws a lead hand punch. To execute the counter to the lead hand punch, Marines—

Step forward and to the left at approximately a 45-degree angle, moving in to the outside of the opponent's attacking arm.

Raise the left arm and block or deflect the opponent's lead hand with the palm of the hand or the meaty portion of the forearm.

Hit and stick by leaving the left arm against the opponent's right arm while stepping forward and to the left at approximately a 45-degree angle to close with the opponent.

Execute a finishing technique, such as a strike or a kick, to the opponent's exposed target areas.

Counter to a Rear Hand Punch. This counter is used when the opponent throws a rear hand punch. To execute the counter to the rear hand punch, Marines—

Step forward and to the left at approximately a 45-degree angle, moving in to the outside of the opponent's attacking arm.

Raise the left arm and block or deflect the opponent's rear hand with the palm of the hand or the meaty portion of the forearm.

Hit and stick by leaving the left arm against the opponent's right arm while stepping forward and to the right at approximately a 45-degree angle to close with the opponent.

Execute a finishing technique, such as a strike or a kick, to the opponent's exposed target areas.

Counters to Kicks

Counter to a Front Kick (Left or Lead Leg). This counter is used when the opponent executes a front kick with his left leg. To execute the counter to a front kick, Marines—

Step forward and to the right at approximately a 45-degree angle, moving in to the outside of the opponent's striking leg.

Lower the left arm and block or deflect the opponent's leg with the palm of the hand or the meaty portion of the forearm.

Hit and stick by leaving the left arm against the opponent's leg while stepping forward and to the left at approximately a 45-degree angle to close with the opponent.

Execute a finishing technique, such as a strike or a kick, to the opponent's exposed target areas.

Counter to a Front Kick (Right or Rear Leg). This counter is used when the opponent executes a front kick with his right leg. To execute the counter to a front kick, Marines—

Step forward and to the left at approximately a 45-degree angle, moving in to the outside of the opponent's striking leg.

Lower the left arm and block or deflect the opponent's leg with the palm of the hand or the meaty portion of the forearm.

Hit and stick by leaving the left arm against the opponent's leg while stepping forward and to the right at approximately a 45-degree angle to close with the opponent.

CHAPTER 5

THROWS

This chapter describes all techniques for a right-handed person. However, all techniques can be executed from either side.

The Marine is depicted in camouflage utilities. The opponent is depicted without camouflage.

Marines use throwing techniques to maintain the tactical advantage and to throw the opponent to the ground during close combat. Throws apply the principles of balance, leverage, timing, and body position to upset an opponent's balance and to gain control by forcing the opponent to the ground. Throwing techniques are effective because they are size- and gender-neutral, and they rely on the momentum and power generated by the opponent rather than the strength or size of the Marine. Marines also execute a throw as a devastating attack against an opponent, possibly causing unconsciousness or broken limbs. When Marines execute throws, they must maintain balance and, simultaneously, prevent the opponent from countering a throw or escaping after being forced to the ground.

1. Turning Throw

Marines use a turning throw to take the opponent to the ground while they remain standing. A turning throw can also be executed from a stationary position. It is particularly effective if Marines and the opponent are wearing gear. To execute the turning throw, Marines—

Grasp the opponent's right wrist with the left hand.

Step forward with the right foot, place it against the outside of the opponent's right foot, and pivot so the back of the heel is next to the middle of the opponent's foot.

Hook the opponent's right arm with the right arm and pinch his arm between the biceps and forearm, touching the opponent with the body.

Pull the opponent's wrist downward, keeping it close to the body.

Pivot to the left on the ball of the foot and continue pulling downward on the opponent's wrist while rotating the wrist outward to off-balance him.

2. Hip Throw

Marines use a hip throw to take the opponent to the ground while they remain standing. A hip throw is particularly effective if the opponent is moving forward or pushing. Marines use the opponent's forward momentum to execute the hip throw. To execute the hip throw, Marines—

Grasp the opponent's right wrist with the left hand.

Step forward with the right foot and place it against the outside of the opponent's right foot.

The back of the heel should be next to the opponent's foot.

Step back with the left foot and rotate on the ball of the foot. The back of the heel is next to the opponent's toe. The knees are bent.

Rotate at the waist and hook the right hand around the back of the opponent's body (anywhere from his waist to his head). The side and hip should be against the opponent.

Rotate the hip against the opponent. The hips must be lower than the opponent's.

Use the right hand to pull the opponent up on the hip to maximize contact.

Pull the opponent's arm across the body and, at the same time, lift the opponent off the ground slightly by bending at the waist, straightening the legs, and rotating the body to the left.

3. Leg Sweep

Marines use a leg sweep to take the opponent to the ground while they remain standing. A leg sweep is particularly effective if the opponent is already off-balanced and moving backward or pulling on the Marine. To execute the leg sweep, Marines—

Grab the opponent's right wrist with the left hand and grab the opponent's left shoulder with the right hand.

Note: Grab the opponent's clothing or gear if his wrist and shoulder cannot be grabbed.

Step forward with the left foot and place it on the outside of the opponent's right foot.

Pull the opponent's wrist downward, close to the body, and push his shoulder backward to off-balance him.

Raise the right knee no higher than the waist.

Kick the foot past the opponent's right leg.

Use the heel of the boot to make contact with the opponent's calf (anywhere from the top of the calf down to the Achilles tendon or on the inside of the calf).

Sweep the opponent to the ground.

CHAPTER 6

CHOKES AND HOLDS

This chapter describes all techniques for a right-handed person. However, all techniques can be executed from either side.

The Marine is depicted in camouflage utilities. The opponent is depicted without camouflage.

WARNING

During training, never execute a choke at full force or full speed and never hold a choke for more than 5 seconds.

When Marines correctly perform a choke, they render an opponent unconscious in as little as 8 to 13 seconds. Chokes are easily performed regardless of size or gender. Marines must know how to apply chokes and how to counter a choke or a hold executed by an opponent.

1. Types of Chokes

There are two types of chokes: an air choke and a blood choke. An air choke closes off the airway to the lungs, thereby preventing oxygen from reaching the heart. A blood choke cuts off the blood flow to the brain. Both types can result in unconsciousness and eventual death for an opponent.

Air Choke

An air choke is performed on the opponent's windpipe (or trachea), cutting off the air to the lungs and heart. If Marines execute the air choke properly, the opponent loses consciousness within 2 to 3 minutes. Due to the length of time it takes to immobilize the opponent, air chokes are not recommended.

Blood Choke

A blood choke is performed on the opponent's carotid artery, which carries oxygen-enriched blood from the heart to the brain. The carotid artery is located on both sides of the neck. If Marines execute a blood choke properly, the opponent will lose consciousness within 8 to 13 seconds. The blood choke is the preferred choke because its intended effect (i.e., the opponent losing consciousness) can be executed quickly, ending the fight.

2. Chokes

Front Choke

Marines execute a front choke when they are facing the opponent. Marines use the opponent's lapels or collar to execute a front choke. To execute the front choke, Marines—

Grab the opponent's right lapel with the right hand, making certain that the knuckles or the back of the hand are against the opponent's carotid artery on the right side of his neck.

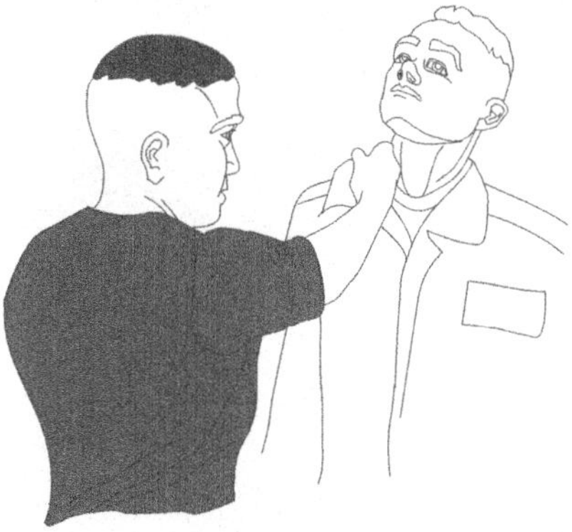

Keep the right hand against the opponent's neck, reach under the opponent's right arm with the left hand, grab the opponent's left lapel, and form an X with the wrists.

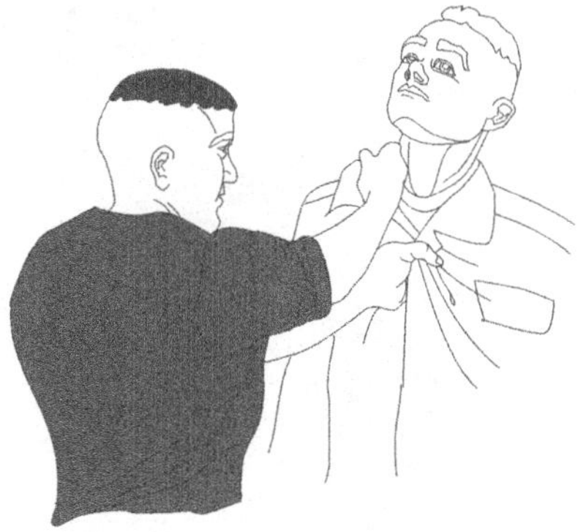

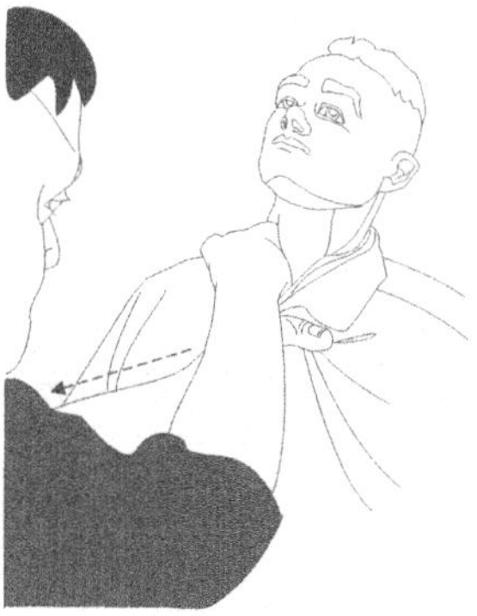

Press the right hand against the opponent's carotid artery.

Pull his lapel to the left with the left hand.

Side Choke

Marines execute a side choke when they are facing the opponent. The side choke is particularly effective when deflecting a punch thrown by an opponent. To execute the side choke, Marines—

Use the left hand to parry the opponent's arm inboard (to the inside of the opponent's reach).

Bring the right arm underneath the opponent's arm and up around the front of his neck.

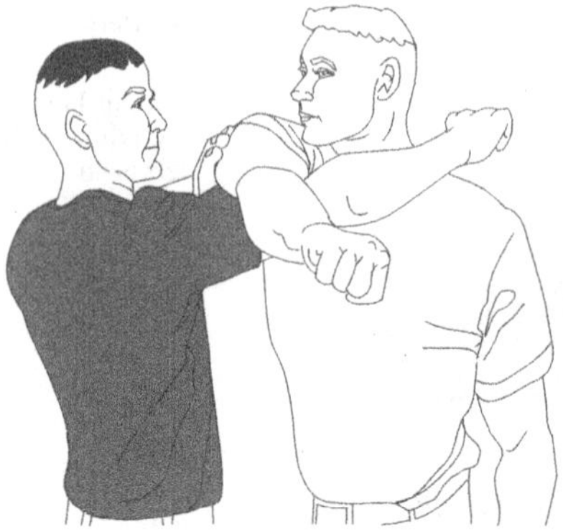

Extend the fingers, place the back of the forearm against the opponent's neck just below his ear, and press the carotid artery on his neck.

Note: If Marines are unable to place their fingers against the opponent's neck, they may use the back of the thumb or wrist.

Reach, with the left hand, around the back of the opponent's neck and clasp the hands together.

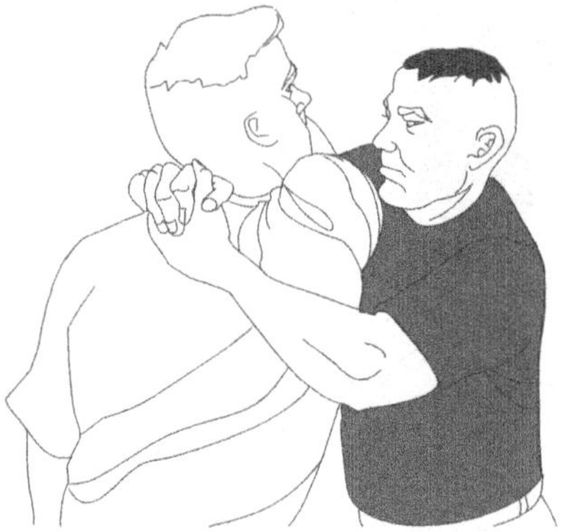

Pull the opponent toward the chest by pulling the clasped hands toward the chest.

Exert pressure on the side of the opponent's neck with the forearm.

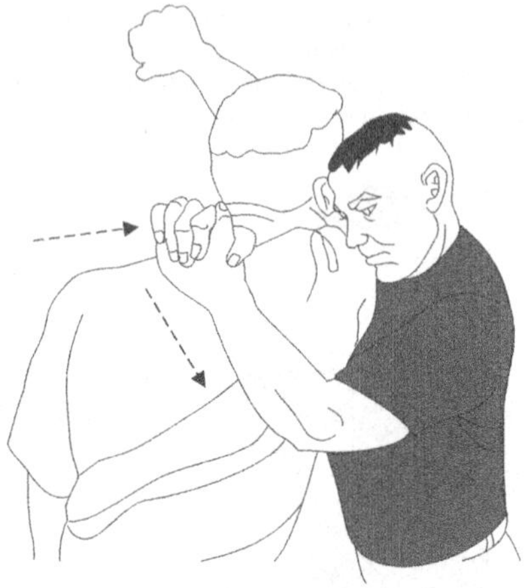

Rear Choke

Marines execute a rear choke when they are behind the opponent, the opponent is on the ground, or when they are taking the opponent to the ground. To execute the rear choke, Marines—

Reach, with the right arm, over the opponent's right shoulder and hook the bend of the arm around his neck.

Clasp both hands together.

Exert pressure with the biceps and forearms on both sides of the opponent's neck on his carotid arteries.

Maintain pressure with the biceps and forearms on both sides of the neck and draw the opponent closer by drawing the right arm in.

Figure-Four Choke

A variation of the rear choke is the figure-four choke. The figure-four choke allows Marines to gain more leverage than the rear choke. If the rear choke cannot be secured, the figure-four variation is employed to increase the pressure of the choke on the opponent. To execute the figure-four variation of the rear choke, Marines—

Apply a rear choke. The Marine's body should be against the opponent's body.

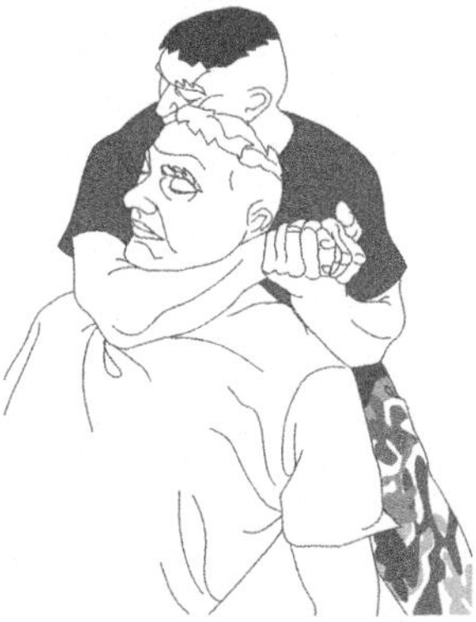

Grasp the left biceps with the right hand and place the left hand against the back of the opponent's head.

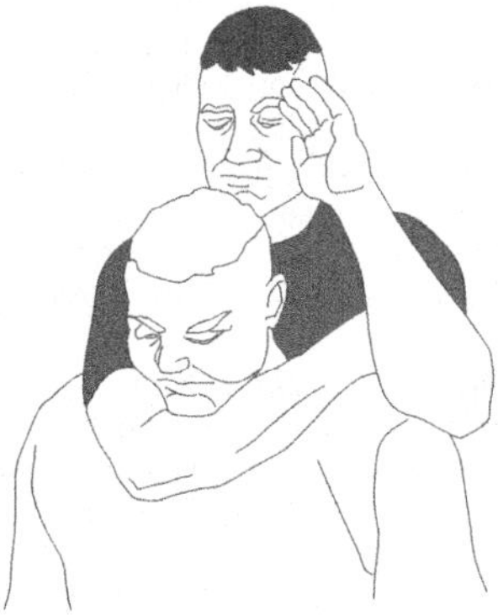

Push the opponent's head forward and down with the left hand.

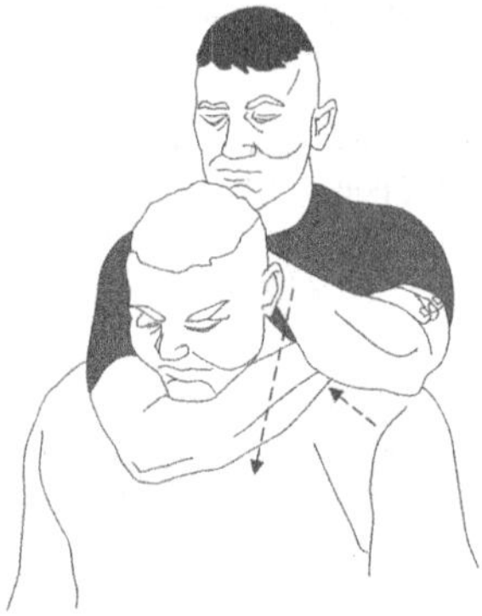

Draw the right arm in, maintaining pressure with the bicep and forearm on both sides of the opponent's neck.

3. Counters to Chokes and Holds

During a close combat situation, an opponent may apply a choke or hold on a Marine. If the opponent correctly applies a choke, a Marine quickly loses consciousness. If a choke is not executed properly, it often results in a hold, typically a bear hug or a headlock. A hold allows the opponent to control a Marine and removes the Marine's ability to attack. It is important for Marines to extract themselves from chokes and holds, regain the tactical advantage, and counter with strikes.

Although a choke causes unconsciousness in 8 to 13 seconds for a blood choke and 2 to 3 minutes for an air choke, the first movement in any counter to a choke is to clear the airway. Marines use softening techniques to loosen an opponent's grip and to clear their airway. Softening techniques are particularly effective if Marines lack the physical strength of their opponent. These techniques include groin strikes, eye gouges, foot stomps, etc. Softening techniques are not offensive; rather, they are used to loosen an opponent's hold.

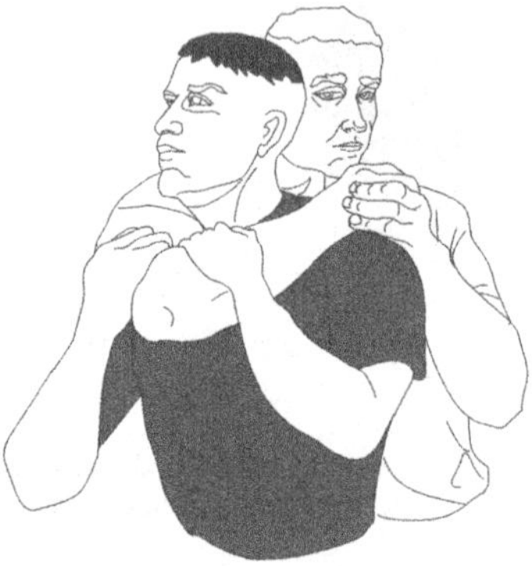

The second movement is to tuck the chin. Once the airway is clear, Marines tuck their chins to prevent the opponent from reapplying the choke.

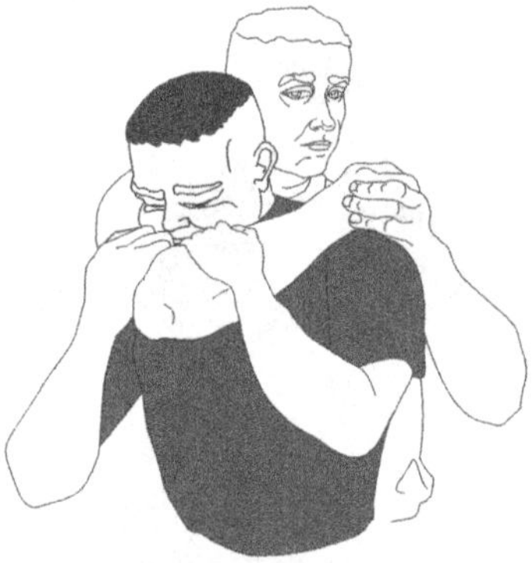

Counter to a Front Choke

Marines use a counter to a front choke when the opponent approaches from the front and uses both hands to choke a Marine around the throat. To execute the counter to the front choke, Marines—

Grasp the opponent's right forearm (where the elbow bends) with the left hand and apply downward pressure on the opponent's radial nerve with the fingers.

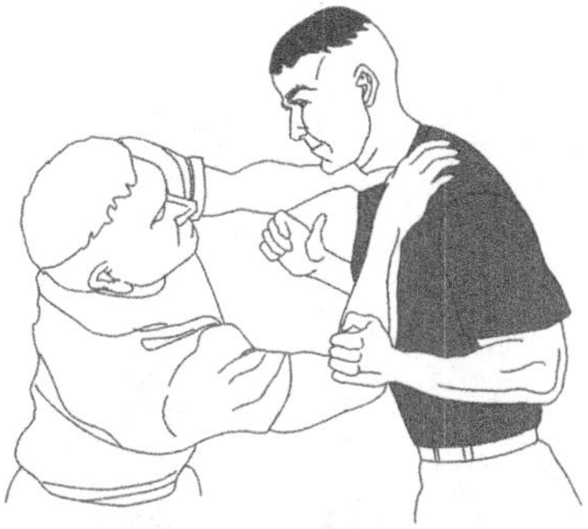

Execute a chin jab to the opponent's chin with the right hand. To generate power into the strike, bring the left foot to the outside of the opponent's right foot and rotate the hips into the strike.

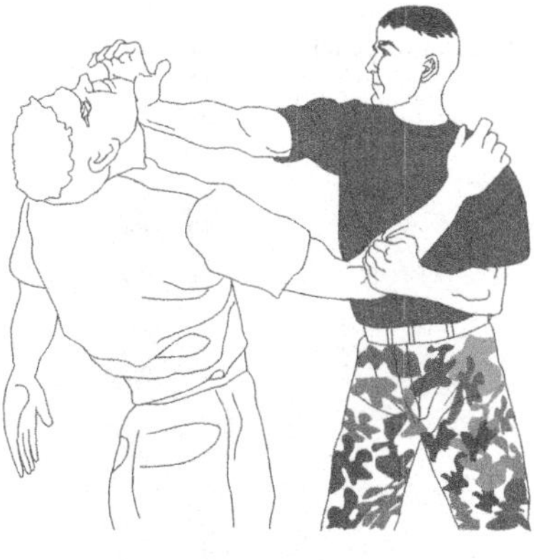

Counter to a Rear Choke

Marines execute a counter to a rear choke when the opponent approaches from the rear and puts his right arm around a Marine's throat. To execute the counter to the rear choke, Marines—

Grasp the opponent's forearm (at the radial nerve) and bicep with both hands and pull down just enough to clear the airway. Once the airway is clear, tuck the chin to protect the airway and to prevent the opponent from reapplying the choke.

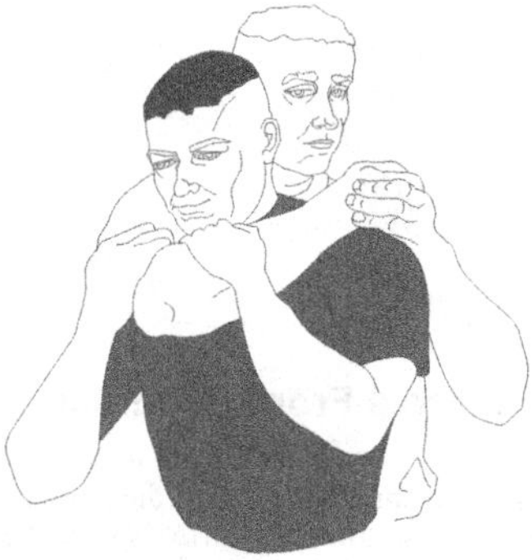

Step behind the opponent's right leg with the left foot, keeping both legs bent (almost in a squatting position).

Strike and drive the left elbow into the opponent's torso while rotating the hips and pivoting to the left, throwing the opponent back and to the ground.

Counter to a Front Headlock

Marines use a counter to a front headlock when the opponent approaches from the front and puts his right arm around the Marine's neck, bends the Marine forward, and locks the Marine's head against his hip. To execute the counter to a front headlock, Marines—

Grasp the opponent's wrist and forearm with both hands and pull down to clear the airway. Maintain control of the opponent's wrist throughout the move. Once the airway is clear, tuck the chin to protect the airway and to prevent the opponent from reapplying the choke.

Move the right hand and arm across the opponent's torso.

Step forward and to the left with the left foot at a 45-degree angle.

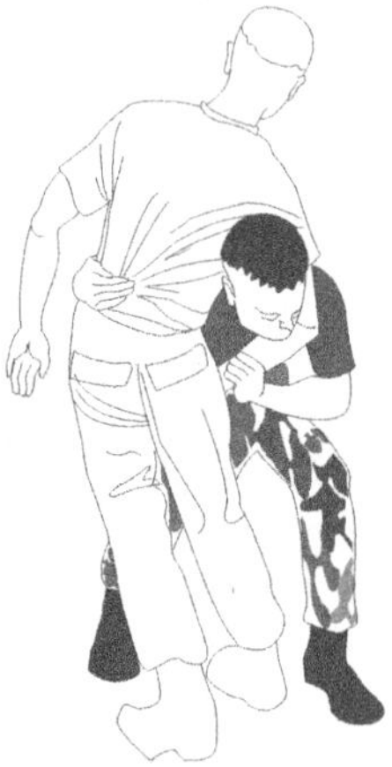

Execute a sweep with the right foot against the opponent's right leg. At the same time, push against the opponent's chest with the right arm and shoulder to generate power in the sweep.

Counter to a Rear Headlock

Marines use a counter to a rear headlock when the opponent approaches from the rear and puts his right arm around the Marine's neck, bends the Marine forward, and locks the Marine's head against his hip. To execute the counter to the rear headlock, Marines—

Grasp the opponent's wrist and forearm with the right hand and pull down to clear the airway. Once the airway is clear, tuck the chin to protect the airway and to prevent the opponent from re-applying the choke.

Reach over the opponent's right shoulder with the left arm.

Grab any part of the opponent's face (chin, nose, eyes) and pull back while rising to a standing position.

Execute, with the right hand, a hammer fist strike to the opponent's exposed throat.

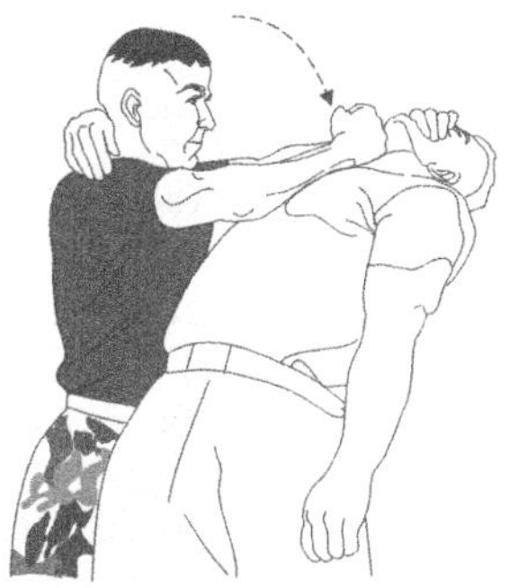

Counter to a Front Bear Hug

Marines execute a counter to a front bear hug when the opponent approaches from the front and puts both of his arms around the Marine's body, trapping the Marine's arms to the sides. To execute the counter to a front bear hug, Marines—

Step forward and to the left with the left foot at a 45-degree angle to the outside of the opponent's right leg, keeping the left leg bent.

Grasp the opponent's torso or arms to gain balance and to assist in throwing the opponent. It may be helpful to hook the opponent's right arm with the left arm.

Drive the right arm and shoulder forward and, at the same time, bring the right leg forward and sweep the opponent's right leg, bringing him to the ground.

Counter to a Rear Bear Hug

Marines use a counter to a rear bear hug when the opponent approaches from the rear and puts both of his arms around the Marine's body, trapping the Marine's arms to the sides. To execute the counter to a rear bear hug, Marines—

Step behind the opponent's right leg with the left foot, keeping both legs bent (almost in a squatting position). The left side of the body should be against the opponent's.

Pivot the hip, turning the body to the left and throwing the opponent back over the bent leg.

CHAPTER 7

GROUND FIGHTING

This chapter describes all techniques for a right-handed person. However, all techniques can be executed from either side.

In drawings, the Marine is depicted in woodland camouflage utilities; the opponent is depicted without camouflage. In photographs, the Marine is depicted in woodland camouflage utilities; the opponent is depicted in desert camouflage utilities.

Marines should avoid being on the ground during a close combat situation because the battlefield may be covered with debris and there is an increased risk of injury. However, many close combat situations involve fighting on the ground. The priority in a ground fight is for Marines to get back on their feet as quickly as possible. In any ground fighting scenario, Marines will usually end up in one of four positions with the opponent. The offensive positions, in which Marines have a tactical advantage, are the guard and mount. The defensive positions, which are used as counters when the opponent has the tactical advantage, are the counter to the guard and the counter to the mount. Marines can also employ chokes during ground fighting to quickly end a fight.

1. Offensive Ground Fighting

Guard Position

Marines execute a guard position when the opponent is on top and the legs are wrapped around the opponent's legs. If the opponent is on top, he may try to choke the Marine, but the Marine still has the tactical advantage because the Marine is in a position to control the situation. To execute the guard position technique, Marines—

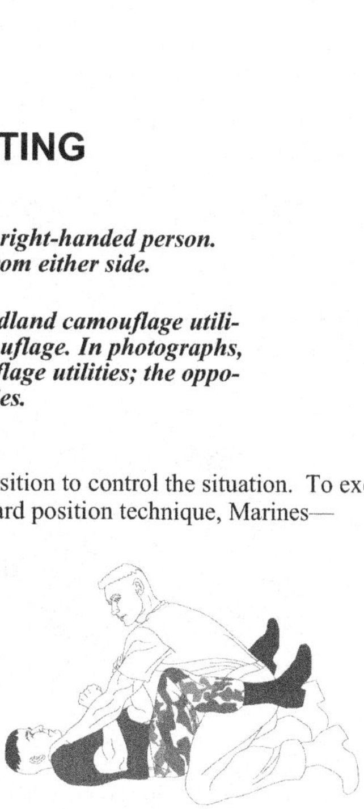

Trap the opponent's hands on the chest by crossing the hands on the chest so the forearms are resting on the opponent's forearms. Apply pressure with the elbows or forearms.

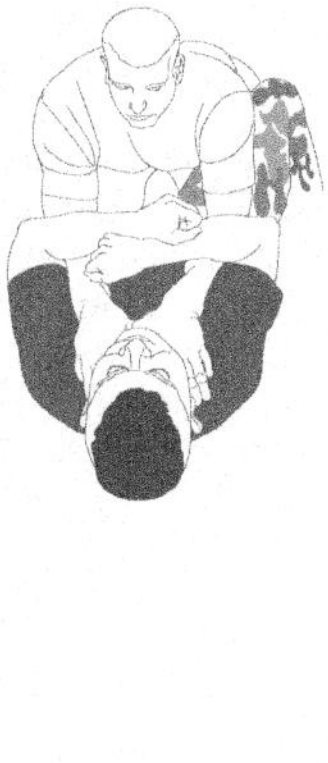

Strike the outside of the opponent's thigh with the cutting edge of the right heel. This causes the opponent to jerk to that side.

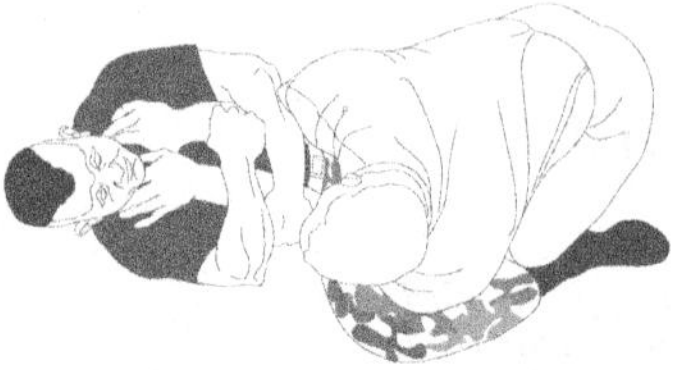

Move the head quickly to the left and swivel the hips to the right. At the same time, bring up both of the legs. Both legs are on the right side of the opponent's body.

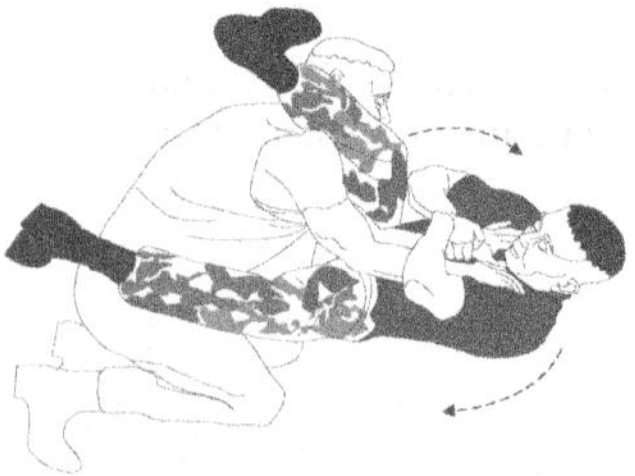

Bring the right leg down, hooking the opponent's neck and head. Exert downward pressure to roll him over on his back. Grasp and maintain control of the opponent's left arm. Upon completion of the move, the Marine is sitting up with legs bent over the opponent while maintaining control of his arm.

Keep legs and knees bent. Maintain pressure against the opponent's neck with the back of the right foot and against his side with the left foot underneath his armpit. Squeeze the knees together, locking the opponent's arm.

Pull the opponent's arm straight up and fall back sharply, pulling his arm to the side in the direction of his little finger.

Return to standing.

Mount Position

Marines execute the mount if the opponent is lying on his back on the ground and the Marine is on top with legs wrapped around the opponent's body. This position is an offensive position because Marines are in a better position to control the opponent and to execute ground fighting techniques. To execute the mount position technique, Marines—

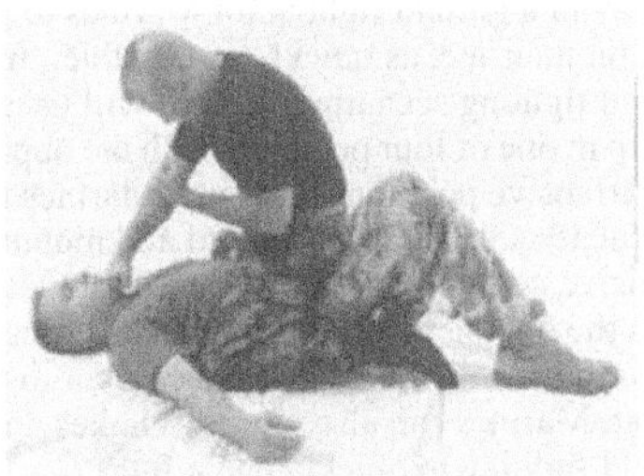

Grab the opponent's wrists or forearms with the hands. Hold them tightly against the torso.

Maintain control of the opponent's arms with the left arm and apply pressure to the opponent's brachial plexus (tie in) with the right hand.

Bring the left foot close to the opponent's right armpit and swing the right leg across the opponent's head.

Keep the legs and knees bent. Maintain pressure against the opponent's neck with the back of the right foot and against his side with the left foot which is underneath his armpit. Squeeze the knees together, locking the opponent's arm.

Pull the opponent's arm straight up and fall back sharply, pulling his arm to the side in the direction of his little finger.

Return to standing.

2. Defensive Ground Fighting

Counter to the Guard

Marines use this technique if the opponent is lying on his back on the ground and the Marine is kneeling on the ground between the opponent's legs. To execute the counter to the guard position, Marines—

Strike the opponent's femoral nerve, located on the inside of the thigh, with the elbows. This forces the opponent to separate his legs.

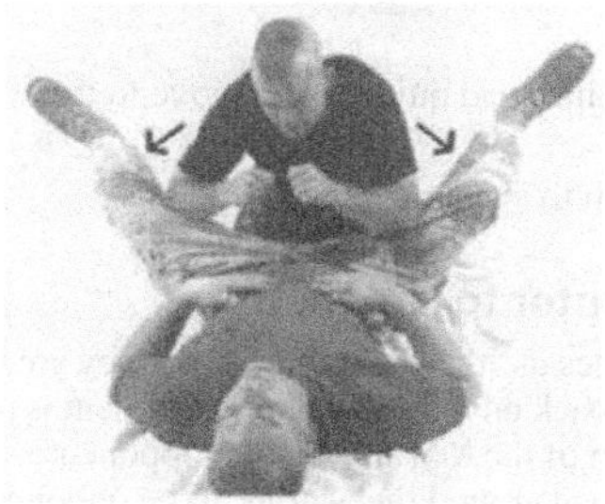

Strike the opponent's groin with the right fist.

Hook the left arm underneath the opponent's right knee from the inside. Throw the opponent's leg over the head.

Duck the head quickly and move to the left.

Return to standing.

Counter to the Mount

Marines use this technique when they are lying on their back on the ground, the opponent is mounted on top of the Marine, and the opponent's legs are wrapped around the Marine. The opponent has the tactical advantage. To execute the counter to the mount position, Marines—

Grab the opponent's gear or clothing on his upper torso and pull him down close.

Use the right arm to hook the opponent's left arm, from the inside around the outside, above his elbow. With the right foot, hook the opponent's left leg or ankle.

Draw the elbow in to bend the opponent's elbow, bringing him down close. The arm must be hooked above the opponent's elbow in order to bend it.

Strike the opponent's side with the left hand.

Push the opponent over and roll him off to the right side.

Maintain control of the opponent's hooked arm and move to a kneeling position.

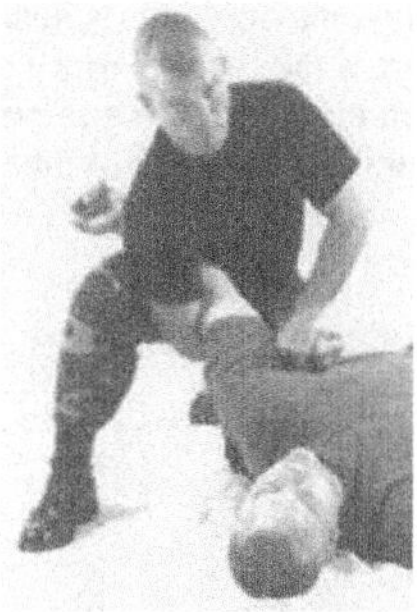

Move to a standing position while maintaining pressure on the opponent's arm and using the knee to apply pressure against the opponent's elbow.

3. Ground Fighting Chokes

WARNING

During training, never execute a choke at full force or full speed. Never hold a choke for more than 5 seconds.

The priority in ground fighting is for Marines to get back on their feet as quickly as possible. Sometimes, Marines can quickly end a ground fight by executing a choke on the opponent. When performed correctly, a choke can render an opponent unconscious in as little as 8 to 13 seconds. Chokes are easily performed regardless of size or gender. The chokes performed during ground fighting are the same as those performed while standing (see chap. 6).

Ground Fighting Front Choke

The ground fighting front choke is a blood choke performed most effectively from the mount position. The front choke employs the opponent's lapels or collar to execute the choke. To execute the ground fighting front choke, Marines—

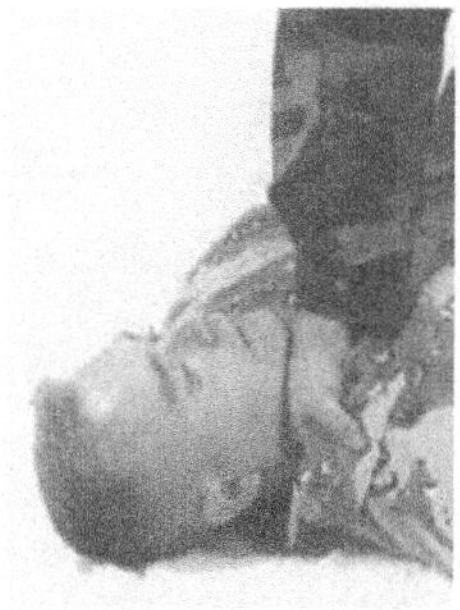

Grab the opponent's right lapel with the right hand, making certain that the knuckles or the back of the hand are against the carotid artery on the right side of the opponent's neck.

Keep the right hand pressed against the opponent's carotid artery, reach under the right arm with the left hand, and grab the opponent's left lapel, forming an X with the wrists.

Keep the right hand pressed against the opponent's carotid artery and pull the opponent's left lapel to the left with the left hand.

Ground Fighting Side Choke

The ground fighting side choke is a blood choke performed from the mount position. The ground fighting side choke is particularly effective when the opponent raises his arms and places them on the Marine's chest or throat. To execute the ground fighting side choke, Marines—

Use the left hand to parry the opponent's right arm inboard (to the inside of the opponent's reach). Bring the right arm underneath the opponent's arm and up around the front of his neck.

Keep the fingers extended, place the back of the forearm against the opponent's neck just below his ear and press the carotid artery on his neck. Depending on the Marine's or the opponent's body configuration, the back of the thumb or wrist may press against the opponent's neck.

Use the left hand to reach around the back of the opponent's neck and clasp the hands together.

Pull the opponent toward the chest. Use the forearm to exert pressure on the side of his neck. This is done by pulling the clasped hands toward the chest.

Ground Fighting Rear Choke

The ground fighting rear choke is a blood choke performed when Marines are behind the opponent. To execute the ground fighting rear choke, Marines—

Wrap the legs around the opponent with the left foot against the inside of his left thigh and the right foot against the inside of his right thigh.

Use the right arm to reach over the opponent's right shoulder and hook the bend of the arm around his neck.

Use the left hand to clasp both hands together.

Exert pressure with the biceps and forearms on both sides of the opponent's neck. This applies pressure to the opponent's carotid arteries. While maintaining pressure with the biceps and forearms, draw the opponent closer by drawing the right arm in. To increase the effectiveness of the choke, lean backward by arching the back and pulling the opponent back. At the same time, push the feet against the opponent's thighs.

Figure-Four Choke

A variation of the ground fighting rear choke is the figure-four choke. The figure-four choke allows Marines to gain more leverage on the rear choke. If the rear choke cannot be secured, the figure-four variation may be applied to increase the pressure of the choke on the opponent. To execute the figure-four variation of the rear choke, Marines—

Apply the ground fighting rear choke. The chest should be against the opponent's back.

Grasp the left biceps with the right hand and place the left hand against the back of the opponent's head.

Use the left hand to push the opponent's head forward and down.

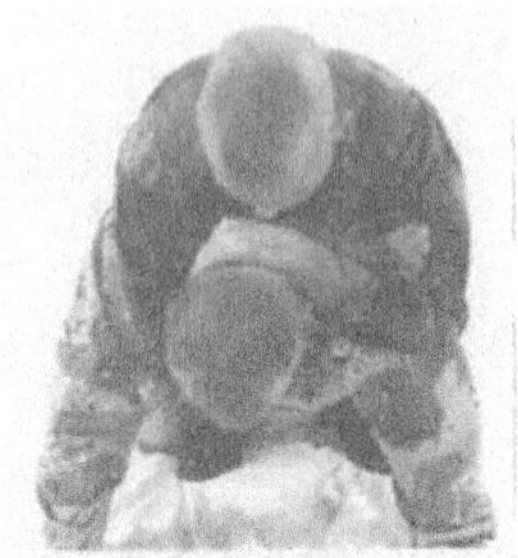

Draw the right arm in while maintaining pressure with the biceps and forearm on both sides of the opponent's neck.

Increase the effectiveness of the choke by arching backward and pushing the feet against the opponent's thighs. Continue to pull the opponent back with the right arm while exerting pressure forward with the left hand against the opponent's head.

CHAPTER 8

NONLETHAL TECHNIQUES

This chapter describes all techniques for a right-handed person. However, all techniques can be executed from either side.

In drawings, the Marine is depicted in woodland camouflage utilities; the opponent is depicted without camouflage. In photographs, the Marine is depicted in woodland camouflage utilities; the opponent is depicted in desert camouflage utilities.

The Marine Corps' involvement in military operations other than war—e.g., humanitarian, peacekeeping, or evacuation missions—has greatly increased. These missions require skills that span the spectrum of conflict and support operations within a continuum of force. But the Corps' day-to-day existence also demands a responsible use of force. Nonlethal techniques are among the skills Marines use to apply a responsible use of force.

1. Unarmed Restraints and Manipulation

Marines operate within a continuum of force, particularly in support of peacekeeping or humanitarian types of missions. In these situations, Marines must act responsibly to handle a situation without resorting to deadly force. Unarmed restraints and manipulation techniques including joint manipulation, come-alongs, and takedowns can be used to control a subject without resorting to deadly force. Marines must train to become proficient in nonlethal techniques and to respond in a responsible manner. These techniques are referred to as compliance techniques, and they are applied in the third level in the continuum of force.

WARNING

During training, never apply the techniques for unarmed restraints and manipulation at full force or full speed. Use a slow and steady pressure to avoid injury.

Compliance Techniques

Compliance techniques are unarmed restraint and manipulation techniques used to physically force a subject or opponent to comply. Compliance can be achieved through the close combat techniques of—

- Pain compliance using joint manipulation and pressure points. (Pain compliance is the initiation of pain to get compliance on the part of the subject.)
- Come-along holds.

Principles of Joint Manipulation

Joint manipulation is used to initiate pain compliance and gain control of a subject. It involves the application of pressure on the joints (elbow, wrist, shoulder, knee, ankle, and fingers). Pressure is applied in two ways:

- In the direction in which the joint will not bend. For example, joints such as the knees and elbows only bend in one direction and when pressure is applied in the opposite direction, pain compliance can be achieved.

- Beyond the point where the joint stops naturally in its range of movement (i.e., it no longer bends).

Since each joint has a breaking point, Marines should apply slow steady pressure only until pain compliance is reached. Continued pressure will break the joint and may escalate the violence of the situation.

Joint manipulation also uses the principle of off-balancing. A subject can be better controlled when he is knocked off balance.

Wristlocks

A wristlock is a joint manipulation that can be applied in a number of ways to achieve pain compliance. The wrist rotates in a number of directions and will bend in a single direction until its movement stops naturally. In a wristlock, pressure is exerted beyond that point by bending or twisting the joint. A wristlock is executed when an opponent tries to grab Marines or is successful in grabbing Marines or their equipment. A wristlock can also be performed by Marines if they wish to initiate control of an opponent.

Basic Wristlock. A basic wristlock is executed when Marines grab the opponent's left hand with the right hand. To execute the basic wristlock, Marines—

Use the right hand to grab the opponent's left hand by placing the thumb on the back of the opponent's hand so that the Marine's knuckles are facing to the left.

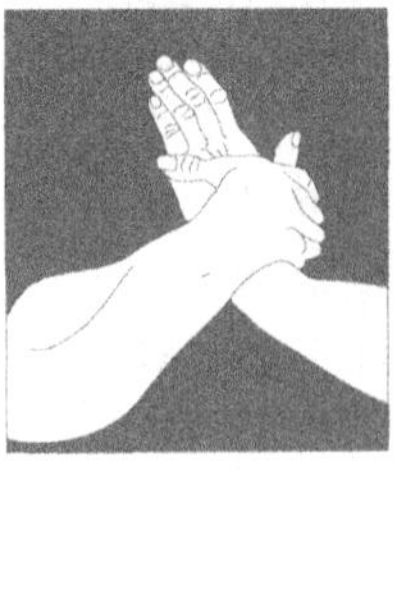

Hook the fingers across the fleshy part of the opponent's palm below the thumb. The fingers are used to anchor the hand so leverage can be applied to twist and bend the joint.

Exert downward pressure with the thumb to bend the opponent's joint. Rotate the opponent's hand to the right to twist the joint.

Use the left hand to further control the opponent.

Step in to the opponent to keep the opponent's hand in close to the body to control him and provide more leverage on the wristlock.

Note: When executing the basic wristlock with the left hand, the Marine grabs the opponent's hand so that the Marine's knuckles are facing to the right, and then rotates and twists the opponent's hand to the left.

Reverse Wristlock. A reverse wristlock is executed when Marines grab the opponent's right hand with the right hand. To execute the reverse wristlock, Marines—

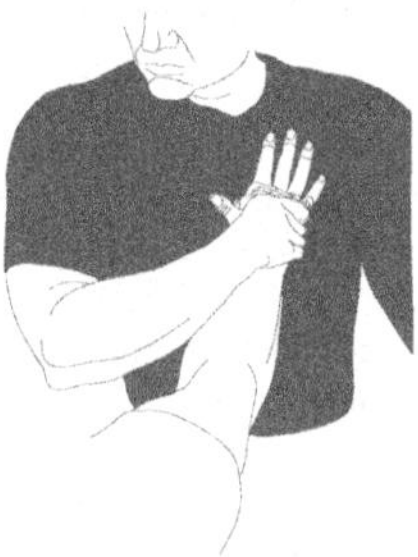

Place the right palm on the back of the opponent's right hand and wrap the fingers across the fleshy part of his palm below his little finger.

Twist the opponent's hand to the right while stepping in to place his hand against the chest. Apply downward pressure on the opponent's hand against the chest. Leave the opponent's hand on the chest to fully control the subject and to gain leverage.

Lean forward to use body weight to add additional pressure to the joint.

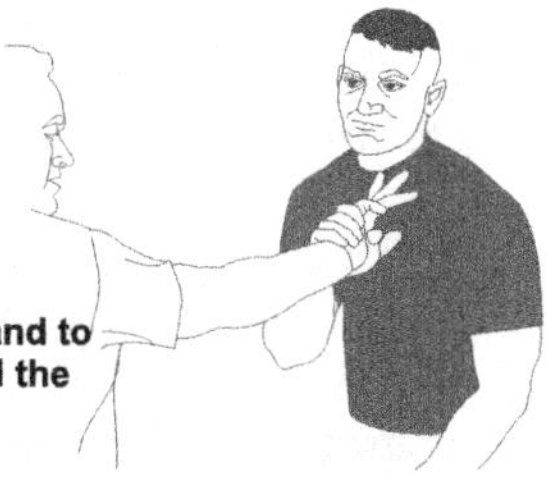

Use the left hand to further control the opponent.

Two-Handed Wristlock. Both hands can be used in the wristlock to maximize the leverage and pressure needed to bend and twist the joint. To execute the two-handed wristlock, Marines—

Place both thumbs on the back of the opponent's hand, thumbs crossed.

Hook the fingers of both hands around the fleshy part of the opponent's palm on both sides of his hand.

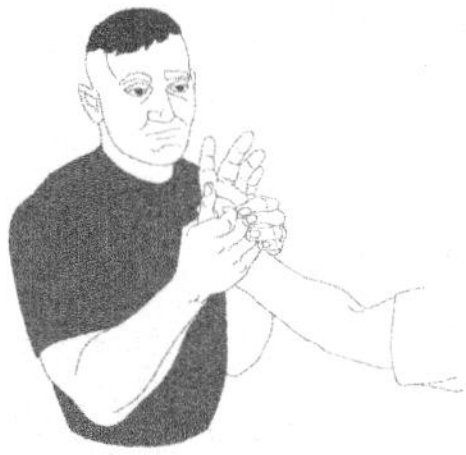

Step into the opponent and apply pressure downward on the back of his hand to bend the joint and rotate his wrist away from the body to twist the joint.

Enhanced Pain Compliance on Wristlock. Enhanced pain compliance techniques are applied in the third and fourth levels in the continuum of force. Additional pain be applied to a wristlock by—

Adding downward pressure to the elbow with the other hand or elbow by using the fingers to pull in on the opponent's radial nerve located on the inside of the forearm. When pressure is added to the opponent's radial nerve, his direction can be controlled.

Applying pressure against the opponent's finger joint to bend it in a direction it cannot bend (i.e., splitting the fingers).

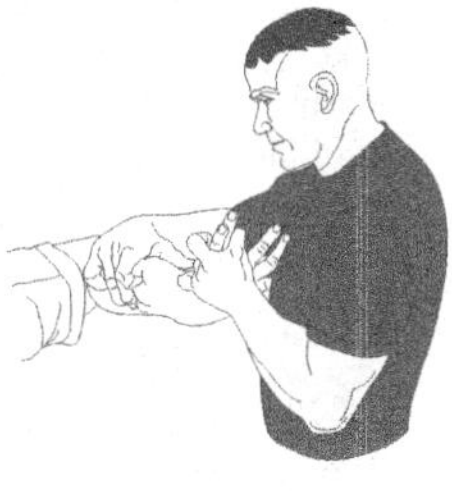

Come-Along Holds

Marines use a come-along hold to control and move an opponent.

Escort Position. A common come-along hold is the escort position. To execute the escort position, Marines—

Face the opponent. Use the left foot to step forward at a 45-degree angle. Turn to face the right side of the opponent.

Use the right hand to firmly grasp the opponent's right wrist. With the left hand, firmly grasp the opponent's right triceps.

Position the opponent's controlled arm diagonally across the torso, keeping his wrist against the right hip. The Marine should be standing to the right of and behind the opponent.

Note: This technique works well when escorting an opponent on either the right or left side. Take caution when escorting an opponent by ensuring his controlled hand is not in a position to grab the holstered weapon. The preferred escort position is from the Marine's left side, so that the opponent is kept further away from the weapon.

Wristlock Come-Along. To execute the wristlock come-along, Marines—

Use the left hand to execute a basic wristlock. Incorporate the right hand in a two-handed wristlock for more control.

Maintain pressure on the opponent's wrist with the right hand, step forward, and pivot around to stand next to the opponent.

Release the left hand, quickly reach under the opponent's arm from behind, and grab his hand.

Use the left hand and apply downward pressure on the opponent's wrist.

Controlling Technique. The following controlling technique is used when an opponent grabs the Marine's wrist. To execute the technique, Marines—

Trap the opponent's hand with the palm of the other hand.

Rotate the opponent's trapped hand up and on his forearm while maintaining downward pressure on his trapped hand.

Apply downward pressure with both hands until the opponent is taken to the ground.

Armbars

An armbar is a joint manipulation in which pressure is applied on a locked elbow, just above the joint, in the direction the joint will not bend. An armbar has to be locked in quickly, but still requires a slow, steady pressure to gain compliance.

Basic Armbar. To execute a basic armbar, Marines—

Use the right hand to grab the opponent's right wrist.

Bring the left hand down on or above the opponent's elbow joint. To gain additional leverage, pivot to face the opponent.

Use the left hand to apply downward pressure on or above the opponent's elbow joint while pulling up on his wrist.

Armbar from a Wristlock. To execute an armbar from a wristlock, Marines—

Use the right hand to grab the opponent's right hand and execute a reverse wristlock.

Bring the left hand down on or above the opponent's elbow joint.

Use the left hand to apply downward pressure on or above the opponent's elbow joint while pulling up on his wrist.

Takedowns

A takedown is used to bring an opponent to the ground to further control him.

Takedown From a Wristlock Come-Along. To take the opponent to the ground from a wristlock come-along, Marines—

Use the right foot to push down on the opponent's calf or Achilles tendon.

Maintain control of the opponent's wrist and elbow and apply a slow, steady pressure to bring him to the ground.

Armbar to a Takedown. This technique is used to take a noncompliant opponent to the ground from an armbar. To execute the armbar takedown, Marines—

Use the right hand to execute a reverse wristlock.

Bring the left hand or forearm down on or above the opponent's elbow joint.

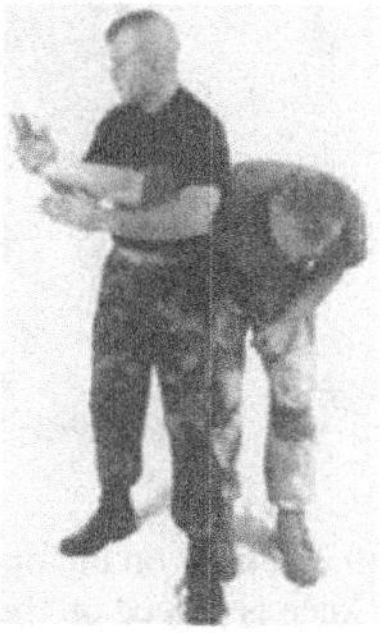

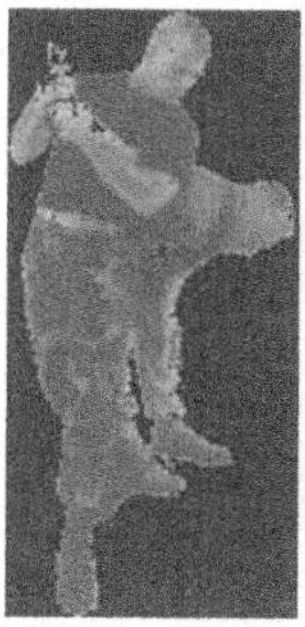

Pivot so the back is facing the opponent and, at the same time, lift the left elbow and slide the body so it is against the opponent, placing the armpit high above the opponent's elbow joint.

Lean back, placing the body weight on the opponent's arm until he complies or is taken to the ground.

Note: This technique may break the opponent's arm. Therefore, this technique should not be employed if the objective is a nonlethal takedown.

Wristlock Takedown. This technique is used to take a noncompliant opponent to the ground from a basic wristlock and to put him in a position where he can be handcuffed, if necessary. To execute the takedown, Marines—

Use the right hand to execute a basic wristlock. Incorporate the left hand in a two-handed wristlock.

Apply downward pressure on the wristlock, pivot on the ball of the right foot, and quickly turn to the right to take the opponent to the ground.

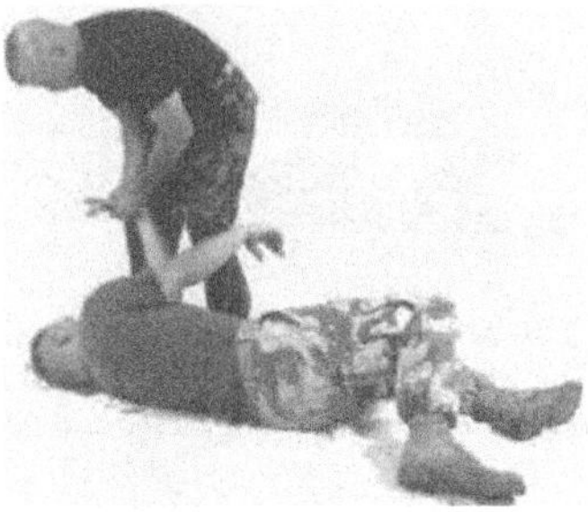

Continue to apply pressure on the wrist joint as the opponent lands on his back with his arm straight in the air.

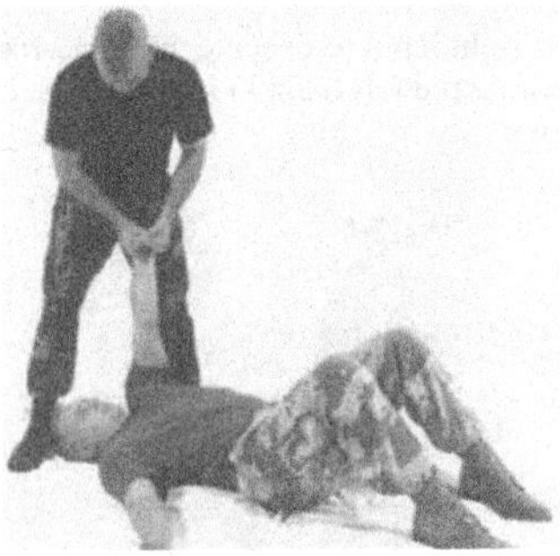

Slide the left foot under the opponent's back, underneath his armpit.

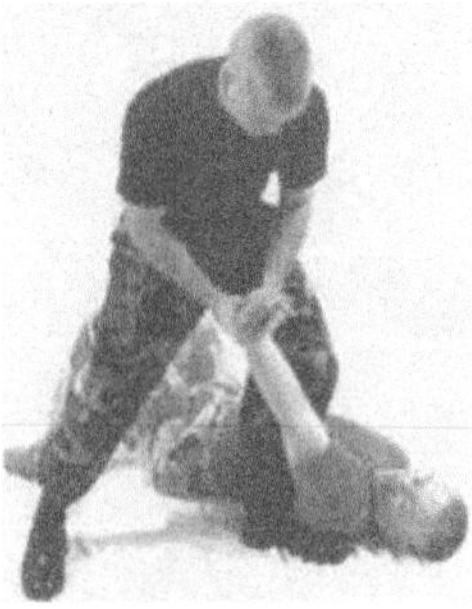

Apply pressure with the knee against the opponent's triceps while pulling back on his arm.

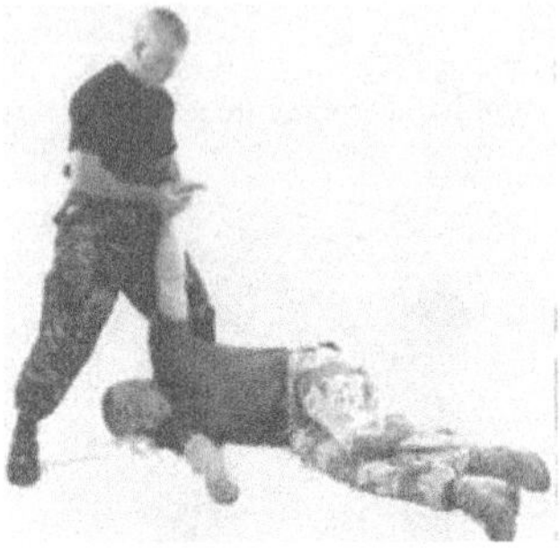

Continue applying pressure with the knee on the opponent's arm. Pivot and step around the opponent's arm to roll him on his stomach.

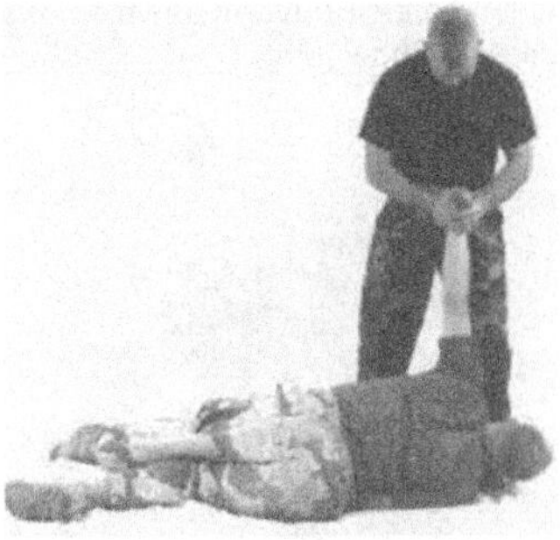

Kneel down with one knee on the opponent's back. The other knee is placed on the opponent's neck and shoulder on either side of his arm. Apply inward pressure with the knees to lock his arm in place.

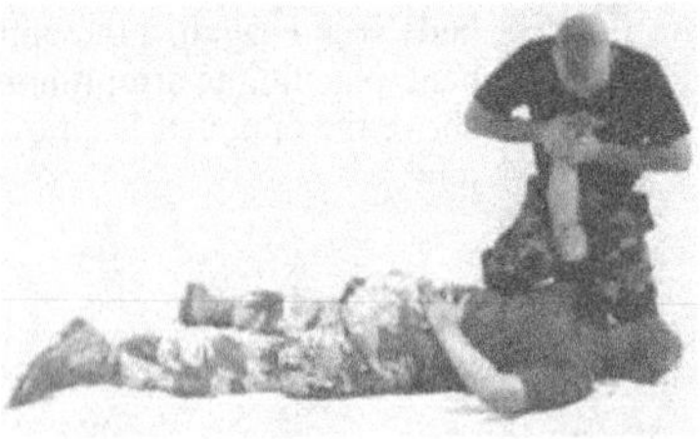

Tell the opponent to put his other hand in the middle of his back. Bring the opponent's controlled hand to the center of his back.

Escort Position Takedown. This technique can be used to control a noncompliant opponent from the escort position. To execute the escort position takedown, Marines—

Lock the opponent's arm straight across the body while rotating his wrist away from the body.

Use the left hand or forearm and apply downward pressure above the opponent's elbow where the triceps meet.

Step back with the right foot and, keeping the opponent's hand controlled against the hip, pivot to the right while continuing to apply downward pressure on his arm to bring him to the ground.

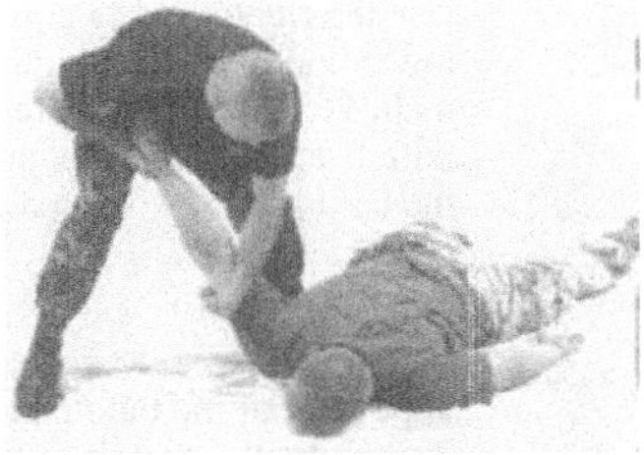

Note: This technique works well when escorting a subject on the Marine's right or left side. When taking down a subject from the right side, step back and pivot to the left.

2. Nonlethal Baton

A baton or nightstick can be an effective compliance tool when used correctly. Batons can be used defensively (blocking), offensively (striking), and as a restraining device when needed. In the fourth level of the continuum of force (assaultive [bodily harm]), defensive tactics include baton or nightstick blocks and blows. Blows to the head or other bony parts of the body are considered deadly force. When deadly force is not authorized, Marines must be able to employ blocks, strikes, and restraints with the baton with the minimum of force.

Grips

One-Handed Grip. To execute the one-handed grip, Marines—

Use the right hand to grasp the lower end of the baton, about 2 inches from the end.

Wrap the thumb and index finger around the baton so they are touching one another. The grip on the baton should be firm, but natural.

Two-Handed Grip. To execute the two-handed grip, Marines—

Use the right hand to grasp the lower end of the baton, about 2 inches from the end. Wrap the thumb and index finger around the baton so they touch one another.

Use the left hand to grasp the upper end of the baton, palm down, about 2 inches from the end. The hands should be approximately 10 to 12 inches apart.

Stance and Method of Carry

The basic warrior stance serves as the foundation for initiating nonlethal baton techniques. The method of carry provides effective defensive positions with a wide range of options to control a combative opponent.

One-Handed Carry. To execute the one-handed carry, Marines—

Grip the baton using the one-handed grip.

Elevate the baton, with the gripping hand at a level between the belt and shoulder.

Keep the left hand in the position of the basic warrior stance.

Two-Handed Carry. This carry is effective for blocks. To execute the two-handed carry, Marines—

Grip the baton using the two-handed grip.

Elevate the baton, with the left hand higher than the right hand.

Orient the weapon toward the subject.

Movement

In a nonlethal confrontation, movement may be made to create distance between Marines and the opponent or to close the gap to control him. When facing an opponent, Marines move in a 45-degree angle to either side of the opponent. Moving at a 45-degree angle is the best way to both avoid an opponent's strike and to put Marines in the best position to control the opponent.

Blocking Techniques

One-handed blocks are used when carrying the baton in the one-handed carry. The same one-handed blocks used in combative stick techniques (see page 3-6) apply in nonlethal baton engagements. The blocks include blocks for a vertical strike, a forward strike, and a reverse strike. Because the baton is often carried with two hands, there are also two-handed blocks that are effectively used from this carry. Two-handed blocks are discussed in the following subparagraphs.

High Block. Marines execute a high block to deter a downward vertical attack directed at the head and shoulders. To execute the high block, Marines—

Raise the baton up to a level even with or above the head. The baton should be in a horizontal position to block the blow. The fingers of the left hand should be open and behind the baton.

Place the baton perpendicular to the opponent's striking surface to absorb the impact of the blow.

Bend the elbows slightly to help absorb the impact of the blow. The arms should give with the strike of the blow.

Low Block. Marines execute a low block to deter an upward vertical attack directed at the abdomen, groin, or torso. The opponent's blow can be delivered by a foot, knee, or fist. To execute the low block, Marines—

Lower the baton to a level even with or below the groin. The baton should be in a horizontal position to block the blow. The fingers of the left hand should be open and behind the baton.

Place the baton perpendicular to the opponent's striking surface to absorb the impact of the blow.

Bend the elbows slightly to help absorb the impact of the blow. The arms should give with the strike of the blow.

Right Block. Marines execute a right block to deter an attack directed at the head, neck, flank, or hip. The opponent's blow can be delivered by a foot, knee, fist, or elbow. To execute the right block, Marines—

Thrust the baton in a vertical position to the right side.

Pivot to the right by stepping with the left foot and pivoting off the ball of the right foot. Rotate the hips and shoulders into the direction of the block. The fingers of the left hand should be open and behind the baton.

Place the baton perpendicular to the opponent's striking surface to absorb the impact of the blow.

Bend the elbows slightly to help absorb the impact of the blow. The arms should give with the strike of the blow.

Left Block. Marines execute a left block to deter an attack directed at the head, neck, flank, or hip. The opponent's blow can be delivered by a foot, knee, fist, or elbow. To execute the left block, Marines—

Thrust the baton in a vertical position to the left side.

Pivot to the left by stepping with the right foot and pivoting off the ball of the left foot. Rotate the hips and shoulders into the direction of the block. The fingers of the left hand should be open and behind the baton.

Place the baton perpendicular to the opponent's striking surface to absorb the impact of the blow.

Bend the elbows slightly to help absorb the impact of the blow. The arms should give with the strike of the blow.

Middle Block. Marines execute either a left or right block to deter an attack directed at the face, throat, chest, or abdomen. To execute the middle block, Marines—

Thrust the baton in a vertical position straight out in front of the body. The fingers of the left hand should be open and behind the baton.

Place the baton perpendicular to the opponent's striking surface to absorb the impact of the blow. The baton should be held with the left hand forward of the right.

Bend the elbows slightly to help absorb the impact of the blow. The arms should give with the strike of the blow.

Restraining Technique

The strong-side armlock is used to restrain an opponent who is not compliant. To execute the strong-side armlock, Marines—

Use the right hand to run the baton up under the opponent's left armpit, parallel to the ground.

Use the right foot to step forward at a 45-degree angle to the left side of the opponent. The baton is across his forearm.

Use the right hand to drive the baton forward and up so the action bends the opponent's arm behind his back. At the same time, continue moving around him to get behind him.

Place the baton on the opponent's forearm with the thumb and/or fingers and apply pressure to his forearm. At the same time, grasp the other end of the baton with the left hand.

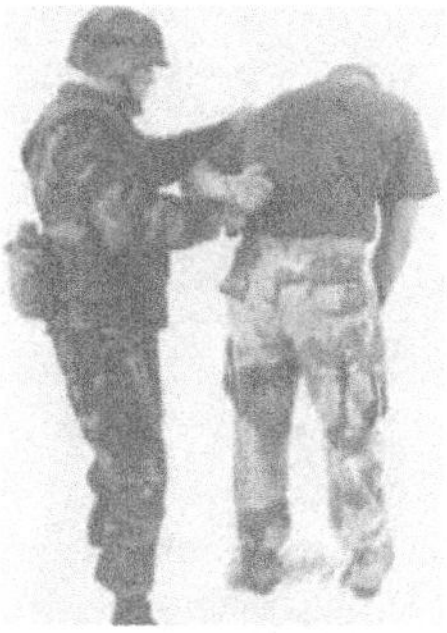

Pull up on the low end of the baton with the right hand. At the same time, push down on the top end of the baton with the left forearm, reaching around with the left hand to grasp the opponent's biceps or shoulder.

Continue exerting downward pressure with the left forearm while pulling back on the opponent's biceps with the left hand. This places the opponent in a position where he is controlled and can be moved.

Apply pressure with the foot against the bend in the opponent's leg above his calf. This lowers the opponent to the ground rather than throwing him to the ground and risking severe injury.

Striking Target Areas

Marines must avoid striking an opponent in the head, neck, or other bony parts with the baton because this is considered deadly force and can lead to serious bodily injury or death. Instead, the legs, arms, and buttocks are target areas that are considered nonlethal.

Legs. Primary targets are the thighs and lower legs. Avoid striking the knee and ankle joints because this can cause permanent damage.

Arms. Primary targets are the upper arms. Avoid striking the shoulder and elbow and wrist joints because this can cause permanent damage.

Buttocks. Primary targets are the buttocks. Avoid striking any other part of the torso, including the chest, rib cage, spine, tail bone, and groin because strikes to these areas can cause permanent damage or death.

One-Handed Striking Techniques

One-Handed Forward Strike. A forward strike follows either a horizontal line or a downward diagonal line using a forehand stroke. To execute the one-handed forward strike, Marines—

Stand facing the opponent with the baton carried in a one-handed carry.

Place the right hand palm up, swing the baton from right to left, and make contact with the opponent.

One-Handed Reverse Strike. A reverse strike follows either a horizontal line or a downward diagonal line using a backhand stroke. To execute the one-handed reverse strike, Marines—

Stand facing the opponent with the baton carried in a one-handed carry.

Bend the right arm and cross the arm to the left side of the body. The baton should be close to or over the left shoulder.

Place the right hand palm down, swing the baton from left to right, and make contact with the opponent.

Two-Handed Striking Techniques

Two-Handed Forward Strike. This strike is an effective follow-up to a middle block or left block. To execute the two-handed forward strike, Marines—

Pull back with the left hand while driving the right hand forward toward the opponent. The baton should be horizontal to the ground. Power is generated by stepping forward with the right foot and rotating the right hip and shoulder into the strike.

Contact the opponent with the end of the baton.

Two-Handed Reverse Strike. This strike is an effective follow-up to a middle block or left block. To execute the two-handed reverse strike, Marines—

Pull back with the right hand while driving the left hand forward toward the opponent. The baton should be horizontal to the ground. Power is generated by stepping forward slightly with the left foot and rotating the left hip and shoulder into the strike.

Contact the subject with the end of the baton.

Front Jab. This strike is effective for countering a frontal attack. It can also be executed as a quick poke to keep a subject away. To execute the front jab, Marines thrust both hands forward in a quick jab. The baton is held either horizontal to the ground or at a slight downward angle.

Rear Jab. This strike is effective for countering a bear hug from the rear. To execute the rear jab, Marines thrust both hands rearward in a quick jab. The baton is held either horizontal to the ground or at a slight downward angle.

APPENDIX A

PUGIL STICK TRAINING

A pugil stick is a training device used to simulate a rifle bayonet so that effective, but safe, training can be conducted to build proficiency of rifle bayonet techniques. Pugil stick training builds on the techniques used to throw punches. Pugil stick training is the only "full contact" training provided to Marines in the Close Combat Program. Pugil stick training teaches Marines to function when faced with stress and violence, and it prepares them to deliver a blow and take a blow. It also provides them with the physical and mental skills vital to success on the battlefield.

1. Pugil Stick Training

Design

A pugil stick consists of a stick wrapped in padding at both ends that can be gripped like a rifle. The pugil stick is approximately the same weight and length of an unloaded rifle with a bayonet attached.

Holding the Pugil Stick

The pugil stick is held in the same manner as the service rifle. All movements come from the basic warrior stance. To hold the pugil stick correctly, Marines—

- Use the right hand to grasp the lower end of the pugil stick overhanded.
- Use the left hand to grasp the upper end of the pugil stick underhanded.
- Use the right forearm to lock the lower end of the pugil stick against the hip.
- Orient the blade end of the pugil stick toward the opponent.

Safety Equipment

The following safety equipment must be worn during any pugil stick bout.

Groin Protection. Groin protection protects the groin from an accidental blow. It should be pulled high around the waist to protect the groin area, with the concave portion against the body.

Flak Jacket. The flak jacket provides protection to the body. It is worn completely fastened.

Neck Roll. The neck roll prevents whiplash if Marines receive a blow to the head. The neck roll further supports the head and protects the neck from blows. The neck roll is worn above the flak jacket and below the helmet. The tied end of the neck roll faces the front.

Helmet. A regulation football helmet protects the face and head. The helmet must fit snugly and the chin straps must be adjusted and snapped.

Mouthpiece. The mouthpiece is worn on the upper teeth to protect the teeth.

Gloves. Marines wear gloves to protect the hands if pugil sticks do not have gloves built-in.

Pugil Stick Screening

Prior to pugil stick training, instructors must ask participating Marines the following questions. If a Marine answers "yes" to any question, he must see the corpsman or a competent medical authority who evaluates him and determines whether or not the he can participate in the training.

- Are you on light duty?
- Are you restricted to running shoes by a corpsman or a doctor?
- Have you fought in a pugil stick bout within the last 7 days?
- Have you received a blow to the head within the last 7 days?
- Have you had a concussion within the last 6 months?
- Have you had dental surgery within the last 24 hours?
- Do you have stitches or staples on your body?
- Have you had a shoulder or head injury within the last 5 years?
- Are you taking a prescription drug?
- Do you have an ear infection or current sinus infection?
- Have you had a broken bone within the last 6 months?

Safety Personnel

The following safety personnel are required to conduct pugil stick training:

- One close combat instructor must officiate the bout.
- One close combat instructor trainer, commissioned officer, or staff noncommissioned officer must be in the training area to serve as range safety officer.
- One corpsman must be in the training area.

Training Area

To prevent injury, Marines train on areas with soft footing (i.e., sand or grass). Training mats are not recommended because feet can stick to the mats, prohibiting movement or causing joint injuries by twisting a knee. Bouts should not take place on a hard surface area; e.g., a flight deck or parking lot. A boxing ring may be used to conduct pugil stick bouts; ring dimensions can vary as long as there is ample room to execute the techniques in the training area.

Second Impact Syndrome

Second Impact Syndrome occurs when a second blow to the head produces a second concussion that occurs within 1 week following a previous concussion (before recovery from the first concussion). Second Impact Syndrome causes rapid brain swelling and can cause death. Therefore, there must be 7 days between pugil stick bouts to reduce the risk of severe injury resulting from Second Impact Syndrome. The 7 day separation between pugil stick bouts significantly reduces the possibility of injury, particularly in someone who may have suffered a brain injury or concussion but shows no symptoms.

Any Marine who experiences headaches or the following symptoms after training must be examined by appropriate medical personnel:

- Blurred vision.
- Ringing in the ears.
- Dilation of the pupils.
- Slurred speech.
- Bleeding from ears or mouth.
- Swelling in head or neck area.
- Any unnatural discoloration in head or neck area.

The Marine should not be allowed to participate in pugil stick training or any other activity where a heavy blow might be sustained for a minimum of 7 days after the headache or other symptoms have subsided.

Safety Measures

The following safety measures must be followed:

- Gear must be worn properly throughout training.
- Contact lenses or glasses will not be worn.
- False teeth will be removed from the mouth.
- Nothing will be worn around the neck except the neck roll.

- Competition among groups of Marines is authorized as long as it does not overshadow training objectives or compromise safety procedures.
- Safety and proper techniques are paramount.
- Safety is more important than competition.

2. General Rules and Regulations Governing Pugil Stick Bouts

Instructors and Support Personnel Requirements

At a minimum, one close combat instructor observes a pugil stick bout. For safety purposes, it is better to have two instructors judging a bout because each instructor can fully observe each of the fighters. The best position for observation is to the right of a fighter. This allows the instructor to see the fighter's facial expression and body movement. The instructor's position must not interfere with the fight.

In addition, one commissioned or staff noncommissioned officer (to serve as range safety officer) and one corpsman will be in the training area.

Prior to the Bout

Prior to the bout, Marines—

- Are paired according to height, weight, and gender.
- Wear the proper safety equipment.

Once Marines have properly donned the safety equipment, they wait for instructions from the close combat instructor. On the command of the close combat instructor, a pair of Marines will enter the ring. Once in the ring, the close combat instructor inspects each Marine for the proper safety equipment.

Note: There should be enough gear so that when two Marines are training, two other Marines can be donning safety equipment for the next bout.

During the Bout

The bout begins when the close combat instructor blows the whistle. All strikes are directed above the waist. Upon hearing a whistle blast, all fighting immediately ceases.

Stopping a Bout

There are two reasons for stopping a bout: delivery of a scoring blow or an unsafe condition. The close combat instructor trainer, close combat instructor, or the range safety officer may stop the bout at any time an unsafe condition is observed.

Scoring Blow. A scoring blow is an offensive technique delivered to a vulnerable area of an opponent with sufficient force and precision to be considered as a disabling or killing blow. Scoring blows are not judged solely on the degree of force with which a blow is delivered, but on the accuracy and techniques employed. A scoring blow is defined as—

- A straight thrust with the blade end of the weapon (red end of the pugil stick) to the opponent's face mask or throat.
- A slash to the side of the opponent's helmet (below the ear) or neck with the red end of the pugil stick.
- A heavy blow to the opponent's head with an authorized technique (i.e., buttstroke, smash) using the butt (black end) of the pugil stick.

When a scoring blow is delivered, the close combat instructor blows the whistle to stop the bout.

Unsafe Condition. The bout will be stopped as soon as an unsafe condition exists. An unsafe condition exists when a Marine is unable to defend himself, loses his balance and falls down on one or two knees or falls down completely, shows instability (e.g., buckling at the knees), loses muscular tension in his neck and his head snaps back or to one side, or appears disoriented. An unsafe condition also exists if a Marine lets go of one end of the pugil stick, the equipment (e.g., helmet, neck roll) falls off, or a Marine fails to use the proper techniques.

If any of these conditions occur—

- The close combat instructor trainer, close combat instructor, range safety officer, or anyone supervising training stops the bout and separates the two Marines.
- The corpsman evaluates the possibly injured Marine to determine if the fight can continue. The corpsman—
  - Ensures that he is alert and responsive, both verbally and physically.
  - Talks to him to see if he is coherent.
  - Makes certain he comprehends and replies to verbal questions.
  - Makes certain his speech is not slurred.
  - Checks his physical signs.
  - Ensures his eyes are focused and not dazed or glazed.
  - Ensures his legs are not wobbly or shaky.

The pugil stick is not used as a baseball bat. The use of unauthorized techniques will result in expulsion from the bout. Only techniques taught in bayonet training are authorized. These techniques include—

- Straight thrust.
- Buttstroke (horizontal and vertical).
- Smash.
- Slash.
- Parry.
- Blocks (low, high, left, right).

After the Bout

A third whistle blast is used prior to identifying the Marine who delivered the killing blow.

A fourth whistle blast is used prior to demonstrating the killing technique used by the Marine who won the bout.

3. Directions for Making Pugil Sticks

Close combat instructor trainers are taught to make pugil sticks to standard specifications at school. It is recommended that all pugil sticks be made under the supervision of a close combat instructor trainer. Until the Marine Corps acquires a standardized pugil stick that units can order through the supply system, instructors use the following information and instructions to make pugil sticks locally.

Supplies

The following supplies are needed to make one pugil stick:

- One 45" by 2" circular, oak, wood dowel.
- Four 9" wide by 27" long foam pad pieces to make two inner handguards of double thickness.
- Four 11" wide by 19" long foam pad pieces to make two outer handguards of double thickness.
- Canvas material to cover inner and outer handguards.
- One foam pad cut 1/4" thick by 7" wide by 13" long to make the centerguard.
- One 8" wide by 14"long piece of canvas material to make a pugil stick sock.
- Canvas material to make two pugil stick end caps. Ideally, two different color fabrics should be used to simulate the bayonet and buttstock ends of a rifle. Common colors for end caps are black and red. Each end cap is a different color to help the officiating certified close combat instructor determine the scoring blow during the pugil stick bout.
- Two 1/4" thick by 11" wide by 54" long foam pad pieces to make two ends.
- Two foam pads cut 1/4" thick by 3" wide by 14 1/2" long to make two end plug inserts.
- One can of 3M Photo Mount™ spray adhesive (NSN 8040-01-120-4009)to adhere foam padding to the stick. (Spray adhesive is available from self-service.)
- One roll of duct tape or riggers tape (NSN 7510-00-266-5016) to protect foam pad and

canvas from wear and to reinforce gluing action of spray adhesive. (Duct tape or riggers tape is available from self-service.)

- One roll para-cord or 5-50 cord to tie both the inner and outer handguards together.
- One roll of thread strong enough to sew canvas material together.
- Sixteen grommets, one for each of the four corners of the inner and outer handguards.
- One foam sleeping mat (NSN 8465-01-109-3369). Use unserviceable foam sleeping mats obtained from Defense Reutilization and Marketing Office.

Tools

The following tools are needed to make a pugil stick:

- Tape measure, ruler, or yardstick.
- Scissors.
- Grommet machine.
- Knife.
- Sewing machine.

Directions

To make a pugil stick, Marines perform the following steps.

Preparing Materials

Step 1. Cut foam sleeping mat into the dimensions listed above. Cut mats to double thickness.

Step 2. Use spray adhesive to adhere the two 9" by 27" pieces of foam together. Use the spray adhesive to adhere the two 11" by 19" pieces of foam together. Let adhesive dry before using the double-thick foam for handguards.

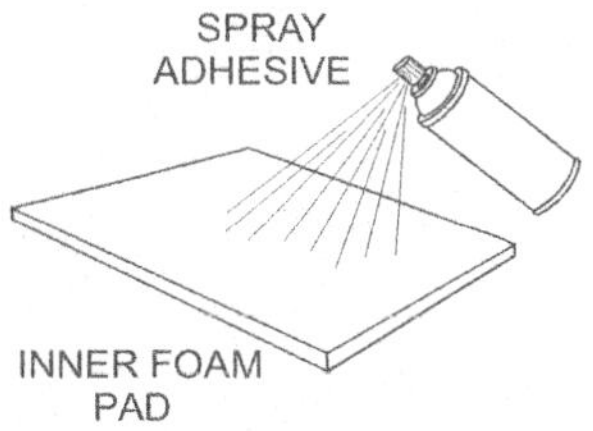

Step 3. Cut canvas material to make two coverings for the inner handguards. Cut material to make two coverings for the outer handguards. Two inner handguards and two outer handguards are required for each pugil stick.

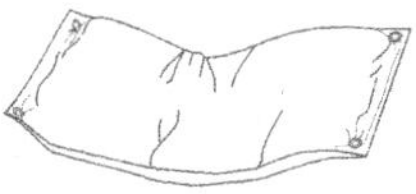

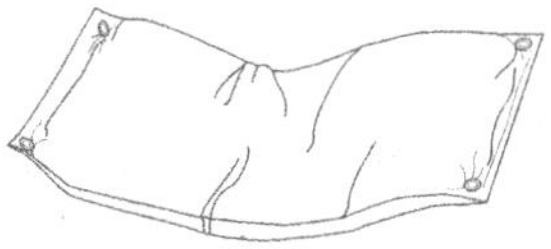

Step 4. Use a sewing machine to make a sleeve out of the 9" by 27" canvas. Sew down three sides. Insert foam pad inside the canvas sleeve and then sew down the fourth side. Use the grommet machine to create holes reinforced with metal grommets in each of the four corners. Once finished, the inner handguards are complete.

Note: It is recommended that the Fabric Repair Shop be used to sew canvas sleeves and apply grommets.

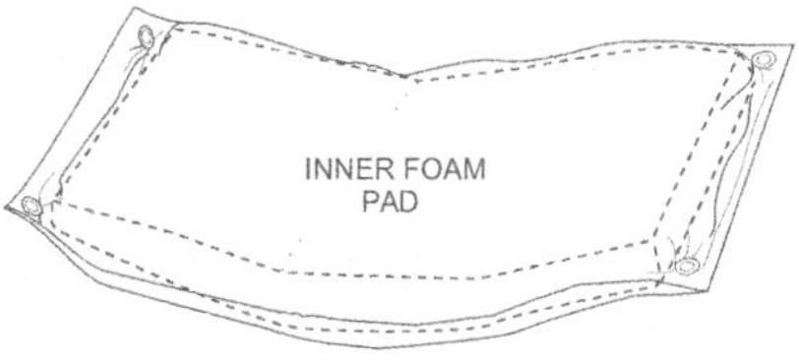

Step 5. Repeat step 4 to make the other inner handguard and both outer handguards.

Constructing Ends, End Plugs, and End Caps

Step 1. Apply spray adhesive on approximately 8 inches of one end of the dowel. This adhesive serves to glue the foam pad to the wooden dowel.

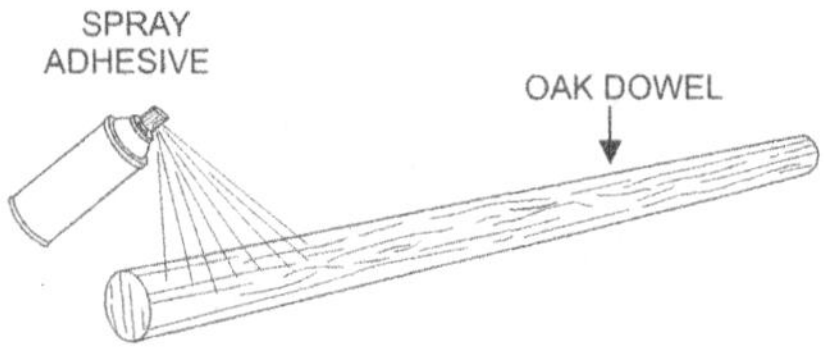

Step 2. Apply spray adhesive on one side of the 11" by 54" foam pad. This causes the foam pad to stick to itself when it is rolled around the end of the wooden dowel. Ensure 3 inches of the rolled foam pad extends past the wooden dowel for stability.

Step 3. Apply spray adhesive on one side of 3" by 14 1/2" foam pad. Roll pad into a tight cylinder.

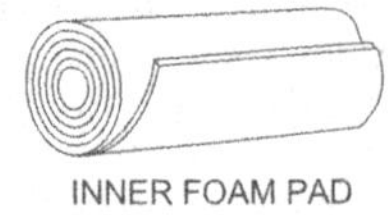

Step 4. Apply spray adhesive on the exposed end of the dowel and fill the hole in the end with the end plug. The end plug prevents the wooden dowel from protruding beyond the edge of the rolled foam padding. This is an important safety precaution to prevent injury.

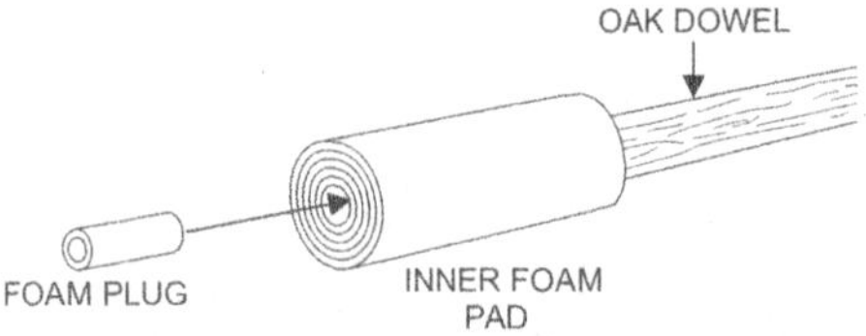

Step 5. Use duct tape or riggers tape to secure and reinforce the foam pad once the end plug is inserted in the end.

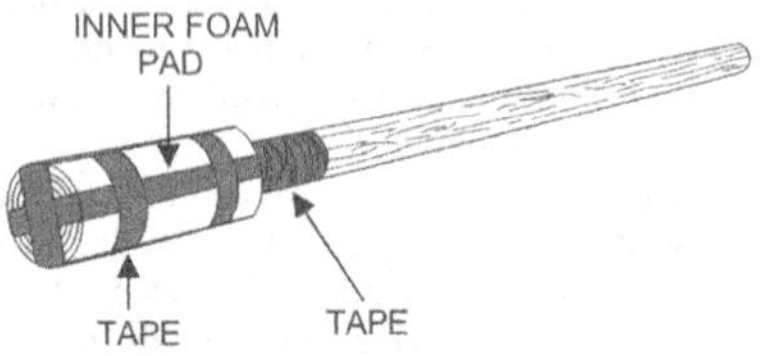

Step 6. Measure the dimensions of the ends (typically, the length of the cylinder should be 12 inches long to allow for taping). Cut and sew canvas material in the shape of a cylinder with one open end. Once complete, this becomes the end cap. The end cap is a colored sleeve that covers the rolled and taped foam pad at each end of the pugil stick.

Note: It is recommended that the Fabric Repair Shop be used to make the end cap.

Step 7. Slide one of the end caps over the rolled foam pad and tape it to the dowel.

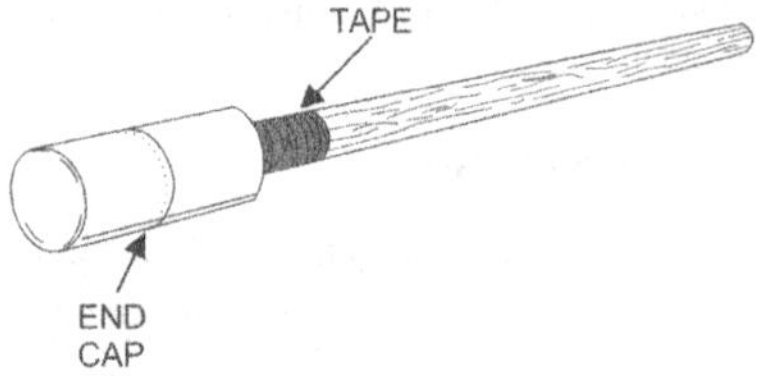

Building the Centerguard

Step 1. Spray one side of the 7" by 13" foam pad with spray adhesive and wrap it tightly around the center of the dowel.

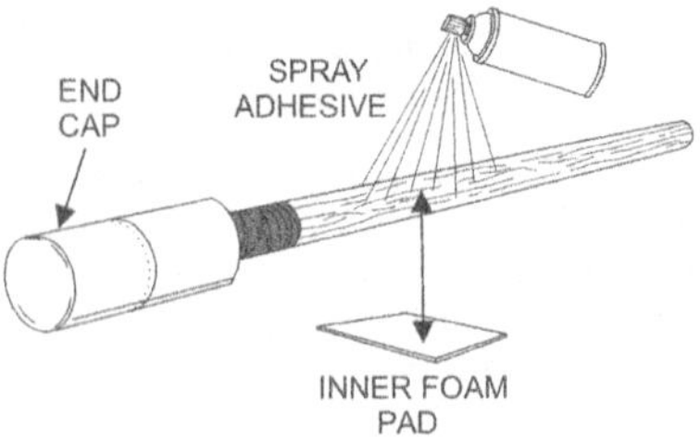

Step 2. Wrap duct tape or riggers tape around the foam pad to secure.

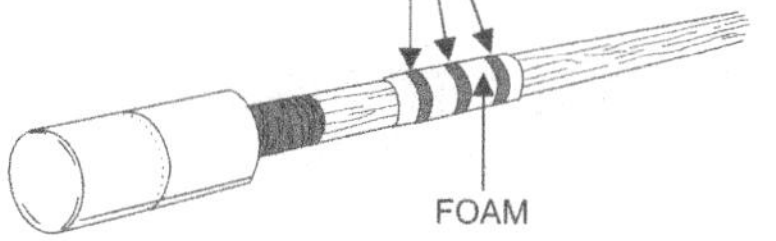

Step 3. Use a sewing machine to make a tubular pugil stick sock out of the 8" by 14" piece of canvas. Slide the pugil stick sock over the rolled foam pad and secure it with the duct tape or riggers tape. Duct tape or riggers tape protects the edges of the centerguard and reinforces the gluing effects of the spray adhesive.

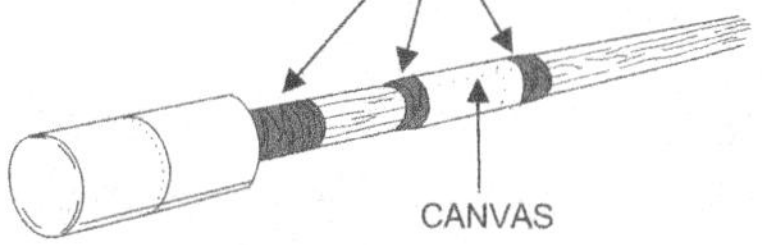

Note: It is recommended that the Fabric Repair Shop be used to make the pugil stick sock.

Finishing the Opposite End Cap

After one end cap is complete and the centerguard is finished, then repeat steps 1 through 7 in Constructing Ends, End Plugs, and End Caps for the other end of the pugil stick. Ensure that the opposite color end cap is used.

Fastening Handguards

Step 1. Place the inner handguards (longer piece) around the dowel. The inner handguard is placed on first because it protects the wrist. One inner handguard is needed for the left hand and one for the right hand.

Step 2. Fold the inner handguard around the pugil stick and run the para-cord or 5-50 cord through the metal grommets.

Step 3. Tighten down the corner ends of the inner handguards by pulling on the para-cord or 5-50 cord.

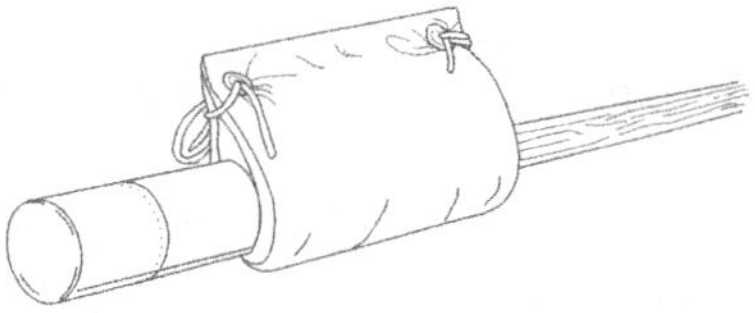

Step 4. Tie off the cord with a square knot above the metal grommets.

Step 5. Place the outer handguards (shorter piece) over the inner handguards (longer piece). Repeat steps 1-4 for the other handguard. The outer handguards provide additional protection to the knuckles and fingers and provides additional stability for the inner handguards.

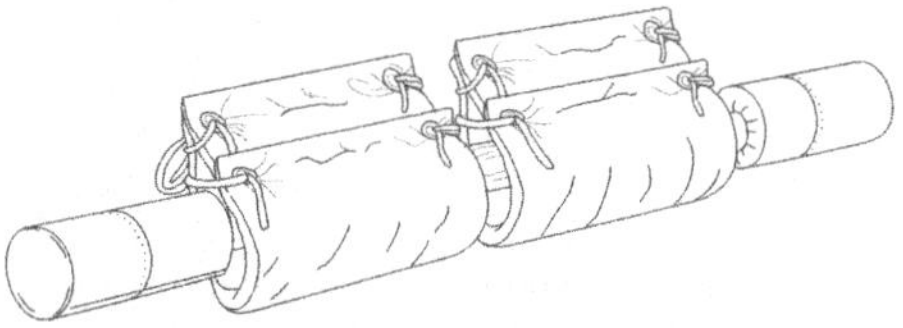

Serviceability Inspection

Prior to any training, all pugil sticks are inspected to ensure they are fit for use. If any of the following conditions exist, the pugil stick is unsafe and will not be used:

- Stick protrudes through the end caps.
- Frayed or worn edges exist on canvas surfaces.
- The stick can be felt through the pads or the padding is too soft.
- End caps or handguards are not securely attached.

APPENDIX B

SAFETY PRECAUTIONS DURING TRAINING

1. General Safety Precautions

When training close combat techniques, certain safety precautions must be adhered to to prevent injuries.

Most training should be conducted on a training area with soft footing such as a sandy or grassy area. If training mats are available, they should be used. A hard surface area is not appropriate for close combat training.

All techniques should be executed slowly at first. Marines can increase the speed of execution as they become more proficient. Marine-on-Marine training that requires contact (chokes, throws, ground fighting, and unarmed restraints and manipulation) should not be executed at full force or full speed.

If a technique is applied to the point that a Marine is uncomfortable, the Marine must "tap out." This indicates immediate release of the pressure being applied or to immediately stop the technique. The Marine "taps out" by firmly tapping his hand several times on any part of the opponent's body that will get his attention or by saying *stop*.

Second Impact Syndrome occurs when a second concussion develops within hours, days, or weeks following a prior concussion (and before recovery from the first concussion). Second Impact Syndrome causes rapid brain swelling and can cause death. Marines who experience headaches or other symptoms following training must be examined by appropriate medical personnel. These symptoms can include, but are not limited to, blurred vision, ringing in the ears, dialation of the pupils, bleeding from the ears or mouth, slurred speech, swelling in head or neck area, or any unnatural discoloration of head or neck. They should not be allowed to participate in pugil stick training or any other activity where a heavy blow might be sustained for a minimum of 7 days after the symptoms have subsided.

2. Safety Precautions for Individual Techniques

Falls

When training Marines to fall, they should progress from the ground, to a kneeling or squatting position, and then to a standing position. This ensures they are comfortable and understand the technique before progressing to executing falls from a higher profile. This instructional technique vastly reduces the risk of injury.

Strikes and Punches

When training Marines to strike and punch, they begin by executing the techniques "in the air." As they become more proficient, they execute strikes on equipment (when available) such as an air shield or a heavy bag. At no time should they be permitted to execute strikes on another student.

Chokes and Ground Fighting Chokes

When training Marines to execute chokes, they will not apply pressure to the opponent's throat during training because the trachea and windpipe can be crushed. During training, Marines should practice the proper procedures for blood chokes. They should never execute air chokes. At no time during training should a choke be applied at full force or full speed. No choke should be held for longer than 5 seconds.

Unarmed Restraints and Manipulation

When training Marines to execute unarmed restraints and manipulation techniques, they utilize slow and steady pressure. Never apply these techniques at full force or at full speed.

Pugil Stick Training

A blow to the head during training will bruise the brain. A second blow to the head can cause death. This is known as the Second Impact Syndrome. There must be 7 days between pugil stick training to prevent injury or death.

PREFACE

1. Scope

This publication describes multi-Service tactics, techniques, and procedures (MTTP) for consideration and use during the tactical employment of nonlethal weapons (NLW) in support of warfighting personnel conducting training and tactical operations. This publication—

a. Provides an overview of NLW and its relationship to deadly force.

b. Provides NLW system description.

c. Describes the capability requirements of NLW.

d. Discusses fundamental concepts and training requirements involved with NLW.

e. Discusses the tactical employment considerations of NLW.

f. Discusses lessons learned from previous NLW use.

2. Purpose

a. This publication provides a single-source, consolidated reference on the tactical employment of NLW and supporting systems.

b. This publication is not intended to restrict the authority of the commander from organizing the force and executing the mission in a manner deemed appropriate.

3. Application

a. This publication provides commanders and their staffs unclassified guidance for NLW employment and planning. Commanders and staffs can use this publication to aid in the tactical employment of NLW during exercises and contingencies.

b. The United States (U.S.) Army, Marine Corps, Navy, Air Force, and Coast Guard approved this multi-Service publication for use.

4. Implementation Plan

Participating service command offices of primary responsibility will review this publication, validate the information and reference, and incorporate it in service manuals, regulations, and curricula follows.

Army. The Army will incorporate the procedures in this publication in U.S. Army training and doctrinal publications as directed by the Commander, U.S. Army Training and Doctrine Command (TRADOC). Distribution of this publication is in accordance with (IAW) initial distribution number (IDN) 115757.

Marine Corps. The Marine Corps will incorporate the procedures in this publication in U.S. Marine Corps doctrinal and training publications as directed by the Commanding General, U.S. Marine Corps Combat Development Command (MCCDC). Distribution of this publication is IAW Marine Corps Doctrinal Publication System (MCPDS).

Navy. The Navy will incorporate the procedures in this publication in U.S. Navy doctrinal and training publications as directed by the Commander, Navy Warfare Development Command (NWDC). Distribution of this publication is IAW Military Standard Requisition-

MARINE CORPS PCN: 143 000057 00
Air Force Distribution: F

ing and Issue Procedures (MILSTRIP) Desk Guide and Navy Standing Operating Procedure (NAVSOP) Pub 409.

Air Force. The Air Force will validate and incorporate appropriate procedures IAW applicable governing procedures. Distribution is IAW Air Force Instruction (AFI) 33-360.

Coast Guard. The Coast Guard will incorporate the procedures in this publication in U.S. Coast Guard doctrinal and training publications as directed by the Assistant Commandant for Operations, Director of Operations Policy, U.S. Coast Guard (USCG). Distribution of this publication is IAW Coast Guard standing operating procedures.

5. User Information

a. TRADOC, MCCDC, NWDC, Air Force Doctrine Center (AFDC), USCG, and the Air Land Sea Application (ALSA) Center developed this publication with the joint participation of the approving service commands. ALSA will review and update this publication as necessary.

b. This publication reflects current joint and service doctrine, command and control organizations, facilities, personnel, responsibilities, and procedures regarding NLW employment. Changes in service protocol, appropriately reflected in joint and service publications, will likewise be incorporated in revisions of this document.

c. We encourage recommended changes for improving this publication. Key comments to the specific page and paragraph and provide rationale for each recommendation directly to—

*FM 3-22.40 (FM 90-40)
*MCWP 3-15.8
*NTTP 3-07.3.2
AFTTP(I) 3-2.45
*USCG PUB 3-07.31

FM 3-22.40 (FM 90-40) — U.S. Army Training and Doctrine Command, Fort Monroe, Virginia

MCWP 3-15.8 — U.S. Marine Corps Combat Development Command, Quantico, Virginia

NTTP 3-07.3.2 — U.S. Navy Warfare Development Command, Newport, Rhode Island

AFTTP(I) 3-2.45 — Headquarters, Air Force Doctrine Center, Maxwell Air Force Base, Alabama

USCG Pub 3-07.31 — U.S. Coast Guard, Washington, D.C.

15 January 2003

NLW

Tactical Employment of Nonlethal Weapons

TABLE OF CONTENTS

Page

*This publication supersedes FM 90-40, MCRP 3-15.8, NWP 3-07.31, USCG Pub 3-07.31, dated 6 October 1998.

FIGURES

Tables

EXECUTIVE SUMMARY

Tactical Employment of Nonlethal Weapons (NLW)

Today's world events mandate a need to project nonlethal force across all levels of war to enable our warfighters and leaders to effectively deal with a host of traditional as well as nontraditional threats. Now more than ever, the minimal level of tolerance for collateral damage and loss of human life, coupled with the tendency for the typical adversary to exploit the rules of engagement (ROE) to his benefit, necessitates an effective and flexible application of force through nonlethal weapons. Much like a rheostat switch in which power can be dialed up or down as desired, NLW provide tools to allow a commander to employ sufficient force to accomplish an objective without requiring the destruction of an enemy or the habitat. The intent of employing NLW is not to add another step in the progression of escalation with an adversary, but to add another tool to use anywhere along that continuum. This publication provides initial guidance for employing NLW in a tactical environment.

Overview of NLW

Chapter I describes the background, definition, and the concept description of NLW. This chapter discusses the development and employment of NLW, and explains the relationship of NLW to deadly force capabilities. The final portion of the chapter reviews policy per Office of Secretary of Defense instructions on employing NLW.

Purpose, System Description, and Organization

Chapter II describes the purpose of NLW and gives a general description of currently available commercial or government off-the-shelf equipment and munitions. The chapter also addresses organization structure, to include system location/distribution, personnel requirements, and mission requirements/tactical structure.

New Capability Requirements

Chapter III describes the applicability of NLW across the range of military operations. Additionally, this chapter discusses the core capabilities as stated in the Joint Mission Area Analysis (JMAA) as described in the Department of Defense Joint Concept for NLW.

Training With the NLW Capability Set

Chapter IV describes training requirements, topics, methods, locations, and a program of instruction from the Interservice Nonlethal Individual Weapons Instructor Course currently available for NLW options. This chapter also addresses additional training available from civilian law enforcement agencies.

Employment of NLW Capabilities

Chapter V establishes the fundamental understanding of the force continuum needed to employ NLW. It outlines employment prerequisites, mission-planning factors, and personnel requirements needed to employ NLW options. It further delineates NLW tasks according to the JMAA and provides considerations for preparation for unit tactical operations. This chapter also discusses employment objectives for NLW and ROE development. This chapter ends with multiple examples of service-specific crowd control formations used by units equipped with NLW.

Lessons Learned

Chapter VI focuses on NLW lessons learned from recent military operations. The chapter discusses how nonlethal technology has changed the way forces are looking at their adversaries in using NLW. It also covers the use of NLW along the force continuum in conjunction with ROE and the media.

PROGRAM PARTICIPANTS

The following commands and agencies participated in developing this publication:

Joint

U.S. Joint Forces Command, Suffolk, VA
U.S. European Command, J3/J5, APO AE
U.S. Pacific Command, J3832, Camp H. M. Smith, HI
U.S. Central Command, J3, MacDill AFB, FL
U.S. Southern Command, SCJ5-PS, Miami, FL
U.S. Space Command, SPJ5X, Peterson AFB, CO
U.S. Special Operations Command, HQ USSOCOM, SOOP-PTE, MacDill AFB, FL
Joint Nonlethal Weapons Directorate, Quantico, VA
Interservice Nonlethal Individual Weapons Instructor Course, Fort Leonard Wood, MO

Army

Headquarters, Department of the Army, ODCSOPS (DAMO-FDT), Pentagon, Washington, D.C.
U.S. Army Training and Doctrine Command, DCSDCS (ATDO-A), Fort Monroe, VA
U.S. Army Training and Doctrine Command, DCSDEV (ATCD-SN), Fort Monroe, VA
U.S. Army Infantry Center, DCD, Fort Benning, GA
U.S. Army Military Police School, ATTN: DOTD, MANSCEN (ATZT-DT-WM-M), Fort Leonard Wood, MO
USAJFKSWCS, DOTD, Fort Bragg, NC
National Ground Intelligence Center, Charlottesville, VA

Marine Corps

Marine Corps Combat Development Command, Quantico, VA
MARCORSYSCOM, Quantico, VA
USMC Warfighting Lab, Quantico, VA
MCSFT Company, MCSF Battalion, Chesapeake, VA
2d FAST MCSF Battalion, Yorktown, VA
1st MARDIV Military Police Company, HQ Battalion, Camp Pendleton, CA
Marine Corps Detachment Military Police and Chemical School, Fort Leonard Wood, MO
Special Operations Training Group, I Marine Expeditionary Force, Camp Pendleton, CA
Special Operations Training Group, II Marine Expeditionary Force, Camp Lejeune, NC

Navy

Commander, Navy Warfare Development Command (NWDC), Newport, RI
USN Surface Warfare Center, Crane, IN
USN Surface Warfare Center, Dahlgren, VA

Air Force

AFDC/DJ, Langley AFB, VA
Headquarters, ACC/SFOT, Langley AFB, VA
Air Force Security Forces Center/SFOR, Lackland, AFB, TX
Headquarters, USAF/XOF, Washington, D.C.

Coast Guard

Commandant, U.S. Coast Guard, (G-OPL-2), Washington, D.C.

Chapter I

Overview of Nonlethal Weapons

1. Background

a. In today's operational environment, United States (U.S.) forces regularly perform missions that were almost inconceivable a few decades ago. In this new environment, lethal firepower or the threat of its use may no longer be the default solution to all crises or problems. Senior leaders face a new level of public and media sensitivity and scrutiny concerning the proper role of the military as an instrument of national power. Field commanders must understand these sensitivities and attempt to achieve an appropriate military force. Junior leaders must apply the resulting decisions wisely, often in changing situations filled with uncertainty and danger.

b. Throughout history, changes in culture and technology have influenced the character of military force and the manner in which it is employed. These changes are an attempt to maximize military force in the force continuum. In the complex and changing political and social domain of the modern world, nonlethal capabilities may offer the opportunity to increase or maximize the military as an instrument of national power.

c. Nonlethal weapons (NLW) must now be considered an integral part of joint operational capabilities. Joint forces anticipate that NLW technology will advance and provide additional options for force commanders.

2. Definition

a. Department of Defense Directive (DODD) 3000.3, Policy for Nonlethal Weapons, 9 July 1996, defines NLW as "weapons that are explicitly designed and primarily employed so as to incapacitate personnel or materiel while minimizing fatalities, permanent injury to personnel, and undesired damage to property and the environment." Furthermore, "unlike conventional lethal weapons that destroy their targets principally through blast, penetration, and fragmentation, nonlethal weapons employ means other than gross physical destruction to prevent the target from functioning. Nonlethal weapons are intended to have relatively reversible effects on personnel and materiel."

b. The term "nonlethal" does not mean zero mortality or nonpermanent damage; these are goals and not guarantees of these weapons. NLW can add flexibility to combat operations and enhance force protection by providing an environment in which friendly troops can engage threatening targets with reduced risk of noncombatant casualties and collateral damage.

3. Concept Description

Nonlethal capabilities within the context of the service capability sets (see Appendix E) may provide armed forces additional offensive and defensive tools to assist in the incapacitation and destruction of an enemy, capture of important intelligence sources, crowd control, civil disturbances, enforcing mandates, and force protection in accordance with the specified rules of engagement (ROE) for a specific operation. These operations include small-scale contingencies (SSC), military operations other than war (MOOTW), stability and support operations (SASO), and civil disturbances. The U.S. Army's Task Force (TF) FALCON in Kosovo and U.S. armed forces located at Camp X-RAY in Guantanamo Bay, Cuba are superb examples of the use of NLW to prevent escalation within an operation.

4. Background on Development and Employment

a. Military forces have long used nonlethal force to influence behavior of people and nations, defeat adversaries with minimum use of lethal capabilities, and weaken adversaries to defeat them with conventional arms. Examples of nonlethal means include a show of force, physical obstacles, riot control agents (RCA), noise to create or enhance psychological effects, smoke and obscurants to mask operations or defeat homing and guidance mechanisms, and using light to disorient combatants. These nonlethal means will remain relevant in future operations and are a part of an evolutionary process of weapons development. Joint operations in Somalia provided increased awareness of the need to incorporate nonlethal options into the fighting capabilities. Therefore, the Department of Defense (DOD) established the Joint Nonlethal Weapons Program in 1996 to research, develop, and acquire these capabilities.

b. Increased interaction between U.S. forces and civilian populations has become a feature of the contemporary operational landscape that is likely to remain the case in the future. Three factors account for this forecast:

(1) Worldwide patterns of population growth and migration have resulted in an expansion of urban culture within established industrialized nations, as well as many pre-industrial and emerging third world societies (60 percent of the world's population live in built-up areas). The prevalence of urbanization in many crisis-prone regions of the world creates the potential for large, vulnerable groups of noncombatants to be entrapped in volatile confrontations involving the deployment of military forces in military operations in urban terrain (MOUT).

(2) U.S. forces increasingly operate in the challenging environment known as MOOTW; a category that includes such missions as humanitarian assistance, disaster relief, noncombatant evacuations, and the various types of peace operations. These operations commonly involve close and continual interaction between U.S. forces and noncombatant civilians. Some MOOTW scenarios include the presence of paramilitary forces, armed factions, or rogue elements that present a constant, but uncertain threat to joint forces. In these situations, the mission of military forces includes preventive tasks. In other words, U.S. forces enhance mission accomplishment by preventing individuals or factions from carrying out specified undesirable activities, such as rioting, looting, attacking, harassing, or threatening. Often, the adversary blends in with the local population of innocent citizens. At other times, sectors of the local population may rise against U.S. forces and become active participants in acts of violence. Factional alignments, the level of violence, and the threat to mission accomplishment, may change frequently with little or no warning. Under such circumstances, the identity of adversaries is uncertain and the use of deadly force for purposes other than self-defense may be constrained by ROE or the judgment of the on-scene commander.

(3) From a defense-in-depth or situational understanding perspective, the force protection, homeland security, and the antiterrorism missions, require that forces assess the use of NLW as a response to potential asymmetric attacks. Asymmetric threats run the spectrum of adversarial use of innovative or nontraditional strategies, tactics, or technologies, to exploit vulnerabilities and strike a decisive blow while avoiding the strengths of an opponent. The objectives of these strikes are intended to undermine national will. Terrorist attacks such as those directed against the USS Cole, World Trade Center, and Pentagon, highlight the severity of the asymmetric threat and its resultant effects.

5. Relationship of NLW to Deadly Force

a. NLW provide "the lesser means" that could be considered prior to employing deadly force. The services definition of deadly force is as follows:

(1) U.S. Army – "Force that a person uses causing, or that a person knows, or should know, will create a substantial risk of causing death or serious bodily harm." (Army Regulation (AR) 190–14, Military Police Carrying of Firearms and Use of Force for Law Enforcement and Security Duties, 12 March 1993).

(2) USMC – "That force which a person uses with the purpose of causing death or serious bodily harm or which a reasonable and prudent person would consider likely to create substantial risk of causing death or serious bodily harm. It will be used as a last resort when all lesser means have failed." Marine Corps Order (MCO) 5500.6F, Arming of Security and Law Enforcement Personnel and the Use of Force, 20 July 1995.

(3) USN – "Force that a person uses causing, or that a person knows or should know would create a substantial risk of causing death or serious bodily harm." Secretary of The Navy Instruction (SECNAVINST) 5500.29B, Use of Deadly Force and the Carrying of Firearms by Personnel of the Department of the Navy in Conjunction with Law Enforcement, Security Duties and Personal Protection, 29 September 1992.

(4) USAF – "Force that causes, or that you know has a substantial risk of causing death or serious bodily harm." Air Force Instruction (AFI) 31-207, Arming and Use of Force by Air Force Personnel, 1 September 1999.

(5) USCG – "Deadly force is any force that is likely to cause death or other serious physical injury." (COMDTINST M16247.1B, Maritime Law Enforcement Manual, 9 August 2001.)

b. The commitment of military power to resolve crises has traditionally involved the use of deadly force or the implicit or explicit threat of the use of deadly force. Military units are trained, organized, and equipped for this purpose. A force armed only with traditional military weapons normally has two options for enforcing compliance: threating deadly force and applying deadly force. This limitation creates a critical vulnerability that belligerents may quickly discern and use to their advantage.

c. Nonlethal capabilities provide a wider range of options that augment, but do not replace traditional means of deadly force. The option to resort to deadly force must always remain available when the commander believes it is appropriate to the mission. DODD 3000.3, states "the availability of NLW shall not limit a commander's inherent authority and obligation to use all necessary means available and to take all appropriate action in self-defense." The existence of nonlethal capabilities does not represent the potential for "nonlethal war," and unrealistic expectations to that effect must be vigorously avoided. Noncombatant casualties, to include serious injuries and fatalities, are often an unavoidable outcome when employing military power, regardless of NLW availability. Augmenting deadly force with NLW should be fundamental to planning and executing any operation where the commander is contemplating deadly force.

d. The commander must clearly articulate the roles of NLW within the ROE. NLW is an additional means of employing force throughout the force continuum for the particular purpose of reducing the probability of death or serious injury to noncombatants or belligerents. Using deadly force must always remain an inherent right of individuals in instances when they, their fellow servicemen, or personnel in their charge are threatened with death or serious bodily harm. NLW add flexibility to combat operations and enhance force protection by providing the field commander with options to engage threatening targets with limited risk of noncombatant casualties and collateral damage.

e. Commanders and public affairs officers (PAO) must prepare personnel to address media questions and concerns regarding the role of NLW. Operational experience indicates that new capabilities give rise to significant media interest. Personnel participating in interviews or briefings must be prepared to address the role of NLW. It must be clear that the presence of NLW in no way indicates abandoning the option to employ deadly force in appropriate circumstances.

6. Advantages of Employing NLW

NLW provide commanders the flexibility to influence the situation favorably with increased safety to friendly forces and reduced risk of both noncombatant fatalities and collateral damage. Some advantages follow:

a. NLW can be more humane and consistent with the political and social implications implicit in humanitarian and peacekeeping missions.

b. The force that properly employs nonlethal options can gain advantages over those who rely on lethal options alone. Because the degree of provocation required to employ NLW is substantially less than lethal options, the force can take a more proactive position that may diminish the likelihood of having a situation escalate to a point where deadly force is required to resolve a conflict.

c. NLW options can be less likely to provoke others. However, commanders should be cautioned using NLW does not preclude an undesired result.

d. Demonstrated restraint can greatly diminish feelings of anger and remorse when deadly force is required after nonlethal options fail.

e. Using NLW can facilitate post incident stabilization by reducing populace alienation and collateral damage.

f. NLW can reduce the possibility of injury to friendly forces compared to forces without NLW.

7. Policy Per Office of the Secretary of Defense Instructions on Employment of NLW

a. DODD 3000.3 states, "nonlethal weapons, doctrine, and concepts of operation shall be designed to reinforce deterrence and expand the range of options available to commanders. Nonlethal weapons should enhance the capability of U.S. Forces to accomplish the following objectives:

(1) Discourage, delay, or prevent hostile actions.

(2) Limit escalation.

(3) Take military action in situations where use of lethal force is not the preferred option.

(4) Better protect our forces.

(5) Temporarily disable equipment, facilities, and personnel.

(6) Nonlethal weapons should also be designed to help decrease the post-conflict costs of reconstruction."

b. DODD 3000.3 continues by stating, "The availability of nonlethal weapons shall not limit a commander's inherent authority and obligation to use all necessary means available and to take all appropriate action in self-defense. Neither the presence nor the potential effect of nonlethal weapons shall constitute an obligation for their employment or a higher standard for employment of force than provided for by applicable law. In all cases, the United States retains the option for immediate use of lethal weapons, when appropriate, consistent with international law."

c. DODD 3000.3 further states, "Nonlethal weapons shall not be required to have a zero probability of producing fatalities or permanent injuries. However, while complete avoidance of these effects is not guaranteed or expected, when properly employed, nonlethal weapons should significantly reduce them as compared with physically destroying the same target." Nonlethal weapons may be used in conjunction with lethal weapon systems to enhance the latter's effectiveness and efficiency in military operations. This shall apply across the range of military operations to include those situations where overwhelming force is employed."

d. According to paragraph 3.21 of the Interium Guidance issued by the Deputy Secretary of Defense, 30 October 2002, all developmental weapon systems must be submitted for legal review and must receive favorable findings before they can be incorporated into the force. This review includes examination for consistency with applicable international and domestic laws, including the laws of war, various arms control treaties, and protocols. U.S. forces can be assured that any NLW they are issued have passed this critical test. Medical and policy reviews should also be considered prior to fielding any new weapon systems. Commanders and operators should consult their specific ROE for guidance.

e. Some capabilities, although designed to minimize fatalities and serious injuries, may have effects that could actually discourage their use. Therefore, the basic decision to employ military force in defense of national interests is usually a matter of intense public concern. The manner in which that force is exercised is subjected to the same scrutiny.

f. The presence of NLW on the battlefield does not mean they must be used. It also does not mean they must be used prior to the use of lethal force. In all cases, the U.S. retains the option for immediate use of lethal weapons when the on-scene commander deems it appropriate. Additionally, the immediate use of lethal weapons must be consistent with U.S. and international law and the ROE.

Chapter II

PURPOSE, SYSTEM DESCRIPTION, AND ORGANIZATION

1. Purpose

The purpose of NLW is to enhance the ability of military commanders to conduct operations across the spectrum of conflict. Over the past several years, military operations have gravitated towards the lower end of this spectrum and have involved U.S. forces in various missions. These missions include SSC, SASO, MOOTW, and civil disturbances where the ROE are inherently restrictive. NLW are designed to address the special requirements of low-intensity environments. They provide tools that enable the individual, team, or unit, to anticipate and/or respond to provocation with more appropriate means than through the use of deadly force. Essentially, NLW provide U.S. forces with a near-term capability for effectively controlling the nontraditional battlefield, within the constraints levied by ROE, by reducing the risk to friendly forces, mitigating casualties, and minimizing collateral damage.

2. System Description

a. NLW Capability Set. The services have, and are continuing to procure, versatile packages of NLW comprised of commercial off-the-shelf (COTS) and government off-the-shelf (GOTS) equipment and munitions. The government is also exploring advanced technologies for future procurements. In addition to addressing contingency requirements, the services' NLW capability sets provide limited sustainment training ammunition and equipment.

b. NLW Capability Set—Component Categories. Each service determines the contents of its own capability set. Examples of the services current configurations are shown in Appendix E. The NLW capability set components can be divided into four distinct categories: personnel protectors, personnel effectors, mission enhancers, and training devices. The following examples are not all-inclusive.

(1) Personnel Protectors. Personnel protectors include items such as face shields, riot shields, and shin guards that protect the individual from blunt trauma injuries inflicted by thrown objects, clubs, etc.

(2) Personnel Effectors. Personnel effectors include items such as riot batons, malodorants, stingball grenades, pepper sprays, and other kinetic rounds (e.g. sponge grenades) designed to discourage, disorient, or incapacitate individuals or groups.

(3) Mission Enhancers. Mission enhancers include items such as bullhorns, spotlights, caltrops, isotropic radiators (stun/flash), and barriers/entanglers. These items are designed to facilitate target identification and crowd control, and to limit personnel and vehicular movement.

(4) Training Devices. Training devices include items such as training suits, training rounds, training batons, and inert pepper sprays. They are designed to facilitate realistic hands-on scenario training in preparation for operations.

c. System Relationships and Inter/Intraoperability. The NLW capability set is designed to provide an operating force (ranging from squad through battalion-sized units [see Appendix E]) with tactically deployable nonlethal equipment. Fielding of this capability set enhances the operation of systems and equipment already fielded and planned within the joint community. The NLW capability set components are compatible with military equipment that is already fielded or planned for future fielding. The components of the NLW capability set are supported within current military logistic capabilities.

d. Augmentation of Organic Equipment. The operational intent of the capability set is to augment an operating force with the required nonlethal equipment to enhance its force projection capability. The NLW capability set is additional equipment that must be incorporated into lift and load planning.

3. Organization

Organization/organizational structure is not anticipated to change as a result of NLW capability sets being added to the inventory.

a. System Location/Distribution. The United States Army (USA), United States Marine Corps (USMC), and United States Navy (USN) have procured and fielded capability sets to the operating forces, security forces, bases, and stations. All services have current plans to field additional capability sets as required. At the time of this printing, the United States Air Force (USAF) has identified its capability sets to field in the near future. The United States Coast Guard (USCG) also intends to field NLW.

b. NLW Capability Set Composition. Appendix E describes the current and proposed capability sets.

c. Personnel Requirements. To date, no change or increase in military specialty codes have been determined in order to field, operate, and maintain the NLW capability set. Each service determines its own training personnel requirements.

d. Mission Requirements/Tactical Structure.

(1) The mission requirements and tactical structure pertinent to the employment of the NLW will vary depending on the type of contingency and subsequent task organization of the employing unit. Mission requirements, derived from commander's guidance, should be satisfied through the appropriate combination of lethal and nonlethal capabilities.

(2) NLW provide commanders with a means of shaping their area of operations (AO) short of employing deadly force. However, it is imperative that U.S. forces retain the means and ability to employ lethal force in the event that nonlethal force fails to achieve desired outcomes. For example, NLW employment may provoke a lethal response from the adversary. Without the imminent threat of lethal force, nonlethal munitions will prove to be less effective and possibly even counterproductive. More importantly, denying U.S. forces a lethal capability would place them at unacceptable risk. If nonlethal force is employed without the threat of lethal force, the belligerent may opt to exploit a perceived vulnerability. In this case, the use of NLW becomes counterproductive, because it serves only to aggravate the threat and causes the situation to deteriorate.

Chapter III

NLW CAPABILITY REQUIREMENTS

1. Applicability Across the Range of Military Operations

a. U.S. forces conduct military operations across a large spectrum. This spectrum spans from humanitarian assistance and peacekeeping to strategic nuclear exchange. Since the end of the Cold War, the U.S. military has been increasingly involved with operations at the lower end of the spectrum. These operations focus on deterring war, resolving conflict, supporting civil authorities and promoting peace. NLW may be useful across many categories of these operations. The range of military operations includes various levels of combat "intensity," and NLW provide one more force option to deal with the varying intensity. MOOTW may be found across the range of military operations and often include situations where noncombatants are mingled with the adversaries or involved in acts of violence, such as rioting or looting. In such circumstances, NLW provide commanders the flexibility to influence the situation favorably with reduced risk of noncombatant fatalities and collateral damage. NLW also provide a means by which to ascertain the intent of individuals/belligerents. Some characteristics of operations that present the opportunity to use NLW are—

(1) Close contact between friendly forces and noncombatant civilians.

(2) The presence of paramilitary or irregular forces.

(3) Armed civilian gangs, unlawful combatants, or militia.

b. These characteristics present a real but ill-defined threat. Under such circumstances, the commander's judgment or ROE may restrict using deadly force for other than self-defense. NLW may provide U.S. forces flexibility to resolve situations where the use of deadly force is restricted. These weapons provide flexibility by—

(1) Allowing U.S. forces to apply measured military force.

(2) Reducing risk of serious noncombatant causalities.

(3) Accomplishing the mission.

(4) Providing force protection.

(5) Retaining the initiative.

(6) Allowing U.S. forces to remain relevant and capable in increasingly complex and chaotic environments.

c. The need to reduce the risks of serious injury to personnel is not limited to crowd control scenarios. In MOUT, some of the local civilian populace may remain in an urban area in the midst of battle. The traditional solution to such challenges has been restricting ROE. Nonlethal capabilities may provide commanders the flexibility to adapt a fluid approach to MOUT, which allows their subordinates the freedom of action to employ appropriate levels of measured military force to accomplish their mission.

d. Another example of NLW applications in MOOTW is in peace enforcement. In these operations, NLW are used in an area denial role to separate belligerents without resorting to deadly force. They are also used to support humanitarian relief organizations in protecting food and medical distribution sites. The presence of NLW will not limit a commander's inherent authority and obligation to use all necessary means available and to take appropriate action in self-defense.

2. Core Capabilities

Core capabilities are those fundamental competencies that enable U.S. forces to achieve desired operational objectives. NLW provide a flexible means to protect friendly forces, high-value assets, and influence the actions of potential adversaries and/or noncombatants. These goals can be achieved without resorting to lethal force and in a manner that may minimize collateral damage. The core capabilities associated with nonlethal effects fall into three major categories as stated within the Joint Mission Area Analysis (JMAA) for Nonlethal Weapons (approved December 2000 by the Joint Requirements Oversight Council): counterpersonnel, countermateriel, and countercapability.

a. Counterpersonnel Capabilities. The DOD Joint Concept for Nonlethal Weapons concept paper, 5 January 1998, describes required nonlethal counterpersonnel capabilities for U.S. forces as follows:

(1) "The application of military force to accomplish a mission with reduced risk of fatalities or serious casualties among noncombatants—or even, in some instances, among enemy forces." There are several specific nonlethal counterpersonnel capabilities to be considered. These include, but are not limited to, the means to influence the behavior and activities of a potentially hostile crowd and the capability to bring a mob engaged in a riot under control. While there are many similarities in these two capabilities, each involves unique challenges, which may require radically different solutions.

(2) "Capability to incapacitate personnel." For the purposes of this document, "incapacitation" is achieved when the weapon's effects render personnel unable to physically or mentally (real or perceived) act in a hostile or threatening manner. In keeping with the guiding principles for NLW, this incapacitation should be reversible through the passage of time. The focus will be on capabilities that affect both individuals and groups of people.

(3) "Capability to deny personnel access to an area (land, sea, or air)." NLW can provide new possibilities for barrier planning in any type of military operation, including systems that produce physical or mental discomfort to those who enter the denied area.

(4) "Capability to clear facilities and structures of personnel." The Joint Concept for NLW states this capability "will facilitate MOUT by reducing the risks of noncombatant casualties and collateral damage while simultaneously minimizing the advantages accruing to an enemy defending a built up area." This capability is applicable to all clearing missions that may be conducted (e.g., Maritime Interception Operations [MIO]) and could assist in providing enhanced force protection.

(5) "Capability to seize personnel, which could include some combination of technologies inherent in other core capabilities, for example, incapacitation and entangling devices (which may aid in seizing personnel)." This capability is intended to augment lethal means used to capture specified individuals, such as enemy combatants or persons who are inciting a mob or crowd to violence.

b. Countermateriel Capabilities. The Joint Concept for Nonlethal Weapons describes countermateriel capabilities as follows:

(1) Rendering equipment and facilities unusable without complete destruction by attacking only weapons of war and supporting infrastructure while minimizing risk to noncombatants. This could enhance operations by defusing potentially volatile situations where conventional military means might otherwise prove counterproductive.

(2) Disabling or denying the use of vehicles, vessels, and aircraft entry into targeted areas or access within an AO. This capability may include physical barriers that impede or stop movement. Alternative systems may alter combustion properties of fuels

and viscosity of lubricants, embrittle or decay materials, attack tires, gaskets, or hoses, or act as adhesives to fuse metal parts. The requirement also includes the capability to facilitate boarding and clearing operations.

c. Counter Capability Capabilities. For U.S. forces counter capability capabilities are described within the JMAA for NLW as follows:

(1) "Disable or neutralize facilities and systems." This includes a nonlethal capability for disabling/neutralizing electrical generating facilities, command, control, communications, computers, intelligence, surveillance, and reconnaissance (C4ISR) systems, Integrated Air Defense Systems (IADS), weapons systems, optical sensors, electrical sensors, and navigation capabilities with such controlled effects as to allow for selective, precise engagement. Other needs include deceiving reconnaissance with illusions of barriers, obstacles, and/or forces, and concealing/covering friendly activities or movements, particularly within complex urban terrain.

(2) "Deny the use of weapons of mass destruction (WMD)." Mission needs include rendering a WMD inoperative; containing the potential release of deadly agents/contaminants; and preventing or neutralizing the production, storage, deployment (transport), employment, and delivery of WMD. Nonlethal capabilities in combating WMD should be designed to permit engagement within a populated or sensitive terrain, prior to the initiation of armed conflict.

3. Limitations

The limitations of NLW are effectiveness, perception, and legality.

a. The effectiveness of NLW is dependent on factors such as motivation, age, environmental and human factors, range of engagement, and numerous others. NLW can be lethal if improperly applied. Furthermore, countermeasures for thwarting virtually all nonlethal options are usually apparent, quickly learned, and readily available. Because they are not intended to kill, nonlethal options teach adversaries what to avoid in the future. Small unit commanders must prepare to keep one step ahead of belligerents.

b. The perceptual limitation is that personnel may misunderstand the appropriate applications of NLW across the range of military operations. The incorrect perception that NLW will allow wars and MOOTW to be prosecuted without casualties may lead to conflicting expectations between political and military leaders. These conflicts can result in U.S. forces being vulnerable to misplaced or inconsistent ROE. All leaders, political and military, involved in planning and executing military missions must understand there are no "nonlethal operations."

c. Some proposed NLW might be forbidden by law or policy. Accordingly, it is essential that all NLW developments be evaluated by appropriate authorities to ensure they comply with the law of war, U.S. law, and U.S. treaty obligations. Use of NLW containing chemical agents, for instance, must be evaluated in the context of the Chemical Weapons Convention. Also, using RCA in an armed conflict requires Presidential approval.

Chapter IV

TRAINING

1. Fundamental Concepts and Training Requirements

The successful accomplishment of any operation in which nonlethal measures are employed requires extensive preparation, of which individual, unit, medical support, and unit training are vital parts. Training should be designed to give individuals an understanding of the entire subject area and enable them to function efficiently as members of a unit. This training must be intensive and realistic. Training with the NLW capability set, and other NLW, is critical to employ these weapons with confidence. The training flow should be—train leaders, train instructors, train unit, exercise. Qualified instructors should conduct initial unit training. Unit sustainment training should be conducted per individual service and unit requirements.

2. Training of Instructors

a. Verbal communication skills, open-hand control techniques, RCA, effects of impact weapons, and civil disturbance tactics, techniques, and procedures (TTP) are areas trained to standard at the Interservice Nonlethal Individual Weapons Instructor Course (INIWIC) held at the Marine Corps Detachment, Fort Leonard Wood, Missouri. INIWIC is an Interservice Training Review Organization (ITRO) course, and quotas are available to any service through normal training reservation channels. A course outline is provided in Appendix C. This is the only formal DOD NLW instructor-training course. The training requirements outlined in the INIWIC represent the minimum training standards to include in developing NLW and tactics instructors. Additionally, the United States Army Military Police School (USAMPS) has developed a training support package that includes Army-specific TTP for small units (company level and below) in using NLW capabilities. The reference for this is the U.S. Army's website at: www.wood.army.mil.

b. The INIWIC is a "train-the-trainer" program to provide commanders NLW instructors within their command. Furthermore, the military services are not precluded from developing their own training based upon the INIWIC curriculum. The commander can then tailor the NLW instruction to fit mission needs. Establishing the service-level INIWIC serves as an assurance measure, giving the commander a quantifiable means of certifying his nonlethal instructors.

3. Nonlethal Training Topics for Deploying Units

Each service and unit will have unique training requirements based on mission type, unit size, deployment area, etc. As a minimum, the training instructor should cover the following topics in any training for units deploying with nonlethal capabilities. They are part of the INIWIC curriculum.

NOTE: U.S. Army training standards are contained in Army Regulation (AR) 350-38, Training Devices Policies and Management, 15 October 1993; USMC training standards for NLW are contained in MCO 1510.112, Individual Training Standards for Nonlethal Weapons, 27 May 1998.

a. Force Continuum. This is an introduction to the federal force continuum model as outlined by MCO 5500.6F, Arming of Law Enforcement and Security Personnel and the Use of Deadly Force; AR 190-14; AFI 31-207, Arming and Use of Force by Air Force Personnel (1 September 1999) (or other service orders). Identifying the proper levels of force, understanding levels of resistance, and how nonlethal technologies affect the force continuum are the minimum topics the instructor should cover.

b. Crowd Dynamics/Crowd Control. This is an introduction to the differences between crowds, mobs, and riots. This course teaches the student basic crowd control techniques that will be easily applied to various situations. The student should be familiarized with tactics and techniques found in field manual (FM) 19-15, MCWP 3-11.1, Rifle Company, TC 90-1, and Air Force manual (AFMAN) 31-201V6, Civil Disturbance (1 April 2002), but will also consider nontraditional and small unit application.

c. Communication Skills. This is an introduction to the stages of conflict management, verbal aggression, nonverbal communication, physical aggression, physiological diversions, and proper mental conditioning. The first NLW skill an individual must have is the ability to communicate. General A. M. Gray, former Commandant of the Marine Corps, once said, "A warrior's most formidable weapon is his mind." The results of an individual's mental processes are evident through actions and words, and the words and communication skills used in dealing with people are truly the first line of nonlethal options.

d. Oleoresin Capsicum (OC) Aerosol and other RCA Training. This is an introduction to the uses of OC/pepper spray and other RCA. The student should have an appreciation for decontamination requirements, storage, and shelf life of OC/RCA. The instructor will also teach legal and tactical considerations. INIWIC certified instructors should teach existing training programs that include live agent training.

e. Open-Hand Control. This is an introduction to pressure point control techniques, unarmed self-defense measures, weapon retention techniques, and other submission/restraint/search techniques. The instructor should cover handcuffing/flex-cuffing for contact team members.

NOTE: USMC plans to replace the current open-hand defensive tactics training with the USMC Martial Arts Program.

---

f. Impact Weapons. This is an introduction to using the rigid straight baton, collapsible straight batons, side handle batons, and/or riot control batons.

g. ROE/Law of Armed Conflict (LOAC). This is an introduction to the ROE/LOAC and their relationship. (Refer to DODD 5100.77, AR 27-1, and NWP 1-14M).

h. Nonlethal Munitions—Employment and Tactics. This subcourse is an introduction to the capabilities and employment of available NLW. NLW provide additional unit force protection capabilities. Introduction to tactics is outlined in Chapter 5. Appendix D is a sample unit training schedule covering the topics listed above. U.S. Army Training and Doctrine Command (TRADOC) has developed training support packages that will include company level and below TTP.

4. Leader Training

a. An understanding of the topics covered in the INIWIC curriculum is essential. However, commanders/leaders should also possess an in-depth understanding of the ROE, their implementation, as well as the politics, culture, and demographics of the area to which they are deploying.

b. Commanders should involve their medical personnel, public affairs (PA) representatives, civil affairs (CA), psychological operations (PSYOP), staff judge advocate

(SJA), engineers, explosive ordnance disposal (EOD), and military police (MP) in training exercises when possible. Involvement by these special staff sections will greatly enhance the quality and realism of the training. Medical personnel will better understand OC decontamination and the possible injuries they may encounter as a result of NLW employment. Because of the international attention that NLW receive, all personnel will benefit from media training by PA personnel. CA and PSYOP personnel can provide area-specific cultural, ethnic, and political briefs to enhance awareness of the challenges within the AO. SJA personnel can clarify complex ROE questions and assist in training on ROE-related issues. Additionally, supply and logistic staff members also require training in NLW maintenance and environmental issues unique to the equipment and munitions.

5. Unit Training

The INIWIC is primarily designed to produce instructors who will train individuals in using and employing NLW. Collective unit tactical training is the unit's responsibility; however, a NLW sample schedule based on INIWIC training is provided in Appendix D. Units should conduct collective training before deploying with NLW. Unit standard operating procedures (SOP) and tactics may require adjustment to enable the employment of NLW and new equipment for the success of specific missions. All collective training should include nonlethal options and decision making. Whenever NLW are employed during training, the instructor must also conduct lethal capability training. Troops must understand the integration of nonlethal and lethal capabilities and always be prepared to protect themselves, as required, with lethal force. It is impossible to accurately predict the response of an individual or crowd.

6. Training Disclaimer

Initial unit training should not be conducted without qualified instructors. Non-type classified and COTS nonlethal munitions cannot be fired without Army Materiel Command (AMC), Army Training Support Center (ATSC), Crane Naval Weapons Center, or Air Force Munitions Center type classifying the ammunition. The instructor should conduct unit sustainment training in accordance with (IAW) service guidelines.

7. Additional Training

a. Additional training is left to the discretion of the unit but should be encouraged in order to add credibility to the training package. Some recommended training certifications for instructors include, but are not limited to—

(1) Management of Aggressive Behavior (MOAB)
R.E.B. Training International, Inc.
PO Box 845, Stoddard, NH 03464

(2) Oleoresin Capsicum Aerosol Training (OCAT)
R.E.B. Training International, Inc.
PO Box 845, Stoddard, NH 03464

(3) Monadnock Defensive Tactics System (MDTS)
Training calendar available at the Monadnock website:
http://www.batons.com/traincal.html

b. These courses add viable credentials to the instructors as perceived by civilian and military alike. In addition, the instructor can issue certification cards for the area which an individual has been certified. Certifications in various areas offers additional protection in the event of a lawsuit or claim against the government by ensuring the individuals that have been trained within your unit are completely capable.

8. Summary

NLW and tactics training are not meant to be stand-alone training—units must anticipate the requirement for NLW in their missions and train accordingly. NLW provide commanders with a graduated response in the force continuum.

Chapter V

EMPLOYMENT

1. Background

a. A military unit deployed to an area will normally conduct operations IAW current doctrine. The environment may contain both noncombatants and combatants. Nonlethal capabilities provide the military force with tools to engage adversaries and minimize collateral damage. Fundamental to employing NLW is a thorough understanding of the force continuum. As stated before, nonlethal capabilities provide a wider range of options that augment traditional means of deadly force but do not replace them. Nonlethal capabilities allow a commander the flexibility to increase and decrease the amount of force applied to accomplish a mission. The force continuum is generally continuous and seamless, yet a careful examination reveals five broad categories:

(1) Threats. The force continuum begins with threats. These can be either implied or expressed. An implied threat may be manifested through presence by forming into riot-control formations, fixing bayonets, or donning protective masks. An expressed threat occurs when a commander makes known the consequences of defiance.

(2) Denial options. The equipment used in denial options are usually placed in highly visible areas and often marked with placards. Examples include concertina, caltrops, barbed wire, or other obstacles.

(3) Munitions that cause physical discomfort/incapacitate. Munitions that cause physical discomfort, but fall short of inflicting trauma, include flash bangs, RCA—tear gas (O-Chlorobenzyl-Malononitrile [CS]), and OC. Although the discomfort or injury may be substantially less than impact munitions, employing these munitions requires an appropriate use of force assessment. Factors such as training, discipline, prejudices, emotions, and judgment all play a part in their application.

(4) Munitions that inflict blunt trauma. Munitions that inflict blunt trauma will inflict painful or debilitating injuries. These munitions constitute the upper end of nonlethal options within the force continuum. Examples include foam batons, stingball grenades, beanbags, sponge projectiles, and rubber ball rounds.

(5) Lethal munitions. Lethal weapons are at the highest level of force continuum. Lethal options should always be regarded as part of the force continuum and not as a separate option altogether. ROE, along with judgment and experience, helps on-scene commanders decide when using lethal force is authorized and appropriate.

b. Nonlethal capabilities include using existing systems that were not designed specifically as NLW. Nonlethal capabilities can minimize fatalities, permanent injury to personnel, and undesired damage to property, facilities, and the environment. Examples include PSYOP, CA, military working dogs, and PA. However, this chapter limits discussion primarily to using weapons and components explicitly designed as part of a NLW capability set. This focus is not intended to limit the commander's option to use whatever means are available to accomplish the mission without unnecessary lethality.

2. Mission Planning Factors

Incorporating NLW into a plan does not change the fundamental principles of mission planning. Commanders should use NLW as an operational enhancement. The first concern of a commander is successfully completing the mission. ROE are a critical consideration in the commander's planning and guide him in determining the appropriate use of nonlethal and lethal force. The availability of NLW does not relieve a force from its obligation to defend itself. Therefore, it must be equipped and trained with both lethal and nonlethal means.

3. NLW Tasks

a. NLW tasks were derived from the combatant commanders and service mission needs statements and are codified in the JMAA for Nonlethal Weapons. These tasks are grouped under three core capabilities: counterpersonnel, countermateriel, and countercapability.

(1) Counterpersonnel:

(a) Control crowds.

(b) Incapacitate personnel.

(c) Deny an area to personnel.

(d) Clear facilities/structures of personnel.

(2) Countermateriel:

(a) Deny an area to vehicles, vessels, and aircraft.

(b) Disable/neutralize vehicles, vessels, aircraft, and equipment.

(3) Countercapability:

(a) Disable or neutralize facilities and systems.

(b) Deny use of WMD.

b. Although not all encompassing, the following list of employment considerations is designed to assist commanders and staffs in formulating plans for use of NLW:

(1) General considerations:

(a) Understand and define ROE.

(b) Commander develops identification of friend or foe (IFF) procedures.

(c) Distribute NLW munitions to troops in advance.

(d) NLW options require lethal overwatch fire.

(e) Ensure area is well lighted to include perimeter.

(f) Employ night vision devices outside the perimeter to identify personnel.

(g) Utilize appropriate media to inform civilians to stay away from denied areas.

(h) Enforce positive access control into perimeter (personnel/vehicles).

(i) Ensure posted signs are easily understood (symbols/pictures). If interpreter support available, post warning signs in native language.

(j) Use entanglements such as caltrops, tanglefoot, barbed wire, and concertina/razor tape as these disrupt/slow intrusion attempts.

(k) Employ sensors, if available.

(l) Employ aerial or naval pickets.

(2) Counterpersonnel considerations:

(a) Employing kinetic rounds at less than 15 feet/5 meters may result in fatal outcomes.

(b) Target area at 15 to 40 feet should be center mass for blunt munitions.

(c) Head shots may be lethal.

(d) Kinetic rounds will not be skip fired.

(e) Kinetic rounds are effective against selected individual targets and can be used to disperse individuals.

(f) Considering differing rates of fires; the M203 and shotgun are complementary in a nonlethal role, depending upon the munitions used and their effective ranges.

(g) Employ riot batons only after proper training.

(h) Use RCA to disrupt/disperse crowds.

(i) Use RCA markers (RCA with a nontoxic, water-soluble marking dye) to mark agitators for future identification and apprehension.

(j) RCA greatly reduce visibility of control force.

(k) Closely control the employment of RCA.

(l) Disperse RCA to face and eyes for full effect.

(m) Employ optics with designated marksmen to identify mob leaders.

(n) Searchlights dazzle/disorient individuals (night only).

(o) Use bullhorn for communicating with crowd and to control formations.

(p) Employ recovery/apprehension/snatch teams to apprehend agitators.

(q) Use flex cuffs for agitators.

(r) Prepare disposition plan for cleared personnel.

(s) Isolate area to prevent reinforcements.

(t) Consider future friendly use of cleared facility.

(u) Prepare security plan for cleared facility.

(v) Channel passive traffic away from formations.

(w) Maintain rapid reaction force/quick reaction force (RRF/QRF).

(3) Countermateriel/countercapability considerations:

(a) Use marking agents to mark a vessel for later interception.

(b) Consider materiel handling equipment required to evacuate vessel/airframe.

c. Selected NLW tasks are further addressed in Tables V-1 through V-7. The tables are not intended to be all-inclusive but merely a stimulus to planning. The equipment listed is taken from the currently available systems designed for nonlethal purposes. The user should not limit his nonlethal options to this specific equipment and should be aware that there are many existing and emerging systems and potential uses of current technology and common objects to achieve the desired mission end state.

Table V-1. Crowd Control

Tasks	***NLW Planning Factors***	***Equipment***	***Employment Considerations***	***Notes***
Crowd control	Administration/ Personnel • SJA / claims planning • Chaplain affairs • Casualty notification • Detainee disposition plan • PAO/media planning • Information dissemination	Support Equipment • Riot face shield • Full length riot shield • Expandable riot baton • Optics • Portable bullhorn • Vest	• Distribute to troops in advance • Employ riot baton only after proper training • Employ optics with designated marksman (DM) to identify (ID) mob leaders • Use bullhorn for communicating with crowd and to control formations	• Strictly define ROE • NLW options require lethal cover fire
Crowd control	Intelligence • Intelligence preparation of the battlespace (IPB) • Background on country and culture • Route/map recon • Sensor emplacement for advance notification • Information on Mob leaders • Patrols and listening post (LP)/observation post (OP) • Determine reason for the crowd	Kinetic Rounds • 12-gauge (ga) bean bag round • 12-gauge rubber bullet • 12-gauge launching cartridge • 40-mm rubber baton • 40-mm wooden baton **NOTE:** USMC no longer uses wooden baton. • 40-mm stinger grenade • Stun grenade • Flash bang	• Do not employ kinetic rounds at less than 15 feet due to possible fatal outcome • Target area at 15 to 40 feet should be center mass • Head shots are not acceptable • Kinetic rounds will not be skip fired • Kinetic rounds effective against selected targets and to disperse individuals • M203 and shotgun are nonlethal complementary; M203 has low rate of fire; area target. Shotgun has high rate of fire; point target	

Table V-1. Crowd Control

Tasks	*NLW Planning Factors*	*Equipment*	*Employment Considerations*	*Notes*
Crowd control	Operations • Tactical Integration of NLW and lethal fires • Fire Support Element (FSE) plans to include RCA • Allocation of NLW munitions • Integration of NLW fires and maneuver • Engineer barrier planning • Understand mission (deny access/disperse/ monitor)	RCA • OC Dispenser • Team OC dispenser • High volume dispenser	• Disrupt/disperse crowds • Disperse RCA to face and eyes for full effect • Employ snatch teams to apprehend agitators	
Crowd control	Logistics • Distribution of NLW supplies • Security of rear area facilities • Resupply of NLW munitions • Procurement of medical supplies • Hospital planning • Medic planning for tactical units	Riot Control • Xenon Searchlight • Flex cuffs • RCA markers	• Searchlights dazzle/disorient individuals (night only) • Flex cuffs used for agitators • RCA markers are used to mark agitators for future ID	

Table V-2. Incapacitate Personnel

Task	*NLW Planning Factors*	*Equipment*	*Employment*	*Notes*
Incapacitate personnel	Administration/Personnel • SJA/claims planning • Media plan	Support Equipment • Riot face shield • Full length riot shield • Expandable riot baton • Rifleman's combat optic • Vest	• Distribute to troops in advance • Do not employ riot baton to the head • Employ optic with DM to ID mob leaders and cover with lethal fire	• Commander needs to define incapacitation mission • Take a balanced approach to public dissemination of NLW so counter measures cannot be rapidly developed • Cover NLW options by lethal force
Incapacitate personnel	Intelligence • ID types of insurgents and demonstrators • ID causes and factions • ID mob ringleaders	Kinetic Rounds • 12-gauge bean bag round • 12-gauge rubber bullet • 12-gauge launching cartridge • 40-mm rubber baton • 40-mm wooden baton **NOTE:** USMC no longer uses wooden baton. • 40-mm stinger grenade • Stun grenade • Flash bang • M203 and shotguns are nonlethal complementary. M203 has a low rate of fire, area target. Shotgun has a high rate of fire, point target	• Do not employ kinetic rounds at less than 15 feet due to possible fatal outcome • Target area at 15-40 feet should be center mass • Head shots are not acceptable • Kinetic rounds will not be skip fired • Kinetic rounds effective against selected targets and to disperse individuals • M203 and shotguns are nonlethal complementary. M203 has a low rate of fire, area target. Shotgun has a high rate of fire, point target	

Table V-2. Incapacitate Personnel

Task	*NLW Planning Factors*	*Equipment*	*Employment*	*Notes*
Incapacitate personnel	Operations • Prepare and confirm ROE for NLW • Integration of NLW and deadly force • Allocate NLW and designate unit use • Integrate NLW fires and maneuver	RCA • OC dispenser • Team OC dispenser • High-volume • OC dispenser	• Disrupt/disperse crowds • Disperse RCA to face and eyes for full effect	
Incapacitate personnel	Logistics • Distribute NLW munitions • ID special storage requirements • Resupply NLW munitions	Riot Control • Xenon • searchlight • Flex cuffs • Chemical markers	• Searchlights dazzle/disorient individuals (night only) • Use flex cuffs for agitators • Use RCA markers to mark agitators for incapacitation	
Incapacitate personnel	Civil Affairs • Collect civilian intelligence • Care for injured personnel • Detain civilian personnel			

Table V-3. Deny Area to Personnel

Task	*NLW Planning Factors*	*Equipment*	*Employment*	*Notes*
Deny area to personnel	Administration/ Personnel • SJA/claims planning • Media plan	Support Equipment • Area lights • Area sensors • Night vision • Access control • Warning signs written in host nation language	• Ensure area is well lighted to include perimeter • Employ night vision devices outside the perimeter • Positively ID accessed personnel • Employ warning signs in native language	• Utilize appropriate media to warn civilians away from denied area

Table V-3. Deny Area to Personnel

Task	*NLW Planning Factors*	*Equipment*	*Employment*	*Notes*
Deny area to personnel	Intelligence • Terrain and map study • Threat ID • Sensor emplacement around area • IPB			
Deny area to personnel	Operations • Prepare and confirm ROE • Integrate NLW and deadly force • Allocate NLW and designate unit use • Determine scope of mission • Determine access procedures • Use of area (friendly or total denial)			
Deny area to personnel	Logistics • Distribute NLW munitions • ID special storage requirements • Resupply NLW munitions • Security of rear area facilities.			
Deny area to personnel	Civil Affairs • Collect civilian intelligence	Entanglements • Caltrops • Tanglefoot • Barbed wire • Foams • Concertina/ razor tape • Stinger spike system	• Disrupt/slow intrusion attempts • Maintain suitable reaction force • Channel passive traffic away • Use foam to cover caltrops	

Table V-4. Deny Area to Vehicles

Task	*NLW Planning Factors*	*Equipment*	*Employment*	*Notes*
Deny area to vehicles	Administration/ Personnel • SJA/claims planning • Media plan	Support Equipment • Area lights • Area sensors • Night vision • Access control • Warning signs written in host nation language	• Ensure area is well lighted to include perimeter • Employ night vision devices outside the perimeter • Positively ID accessed vehicles	• Commander needs to define denial mission • Utilize appropriate media to warn civilians away from denied area
Deny area to vehicles	Intelligence • Terrain and map study • Threat ID • Sensor emplacement around area • IPB • Traffic analysis			
Deny area to vehicles	Operations • Prepare and confirm ROE • Allocate NLW and designate unit use • Determine scope of mission • Determine access procedures • Determine MP/ security forces requirements • Use of area (friendly or total denial)			• Commanders must consider flying operations when employing RCA. Area(s) should be reusable in short order.
Deny area to vehicles	Logistics • Distribute NLW supplies • Be prepared to evacuate stalled vehicles			
Deny area to vehicles	Civil Affairs • Collection of civilian intelligence	Entanglements • Caltrops • Tanglefoot • Barriers • Organic equipment • Barbed wire • Foams	• Disrupt/slow intrusion attempts • Maintain suitable reaction force • Channel passive traffic away • Use foam to cover caltrops	• Commanders must consider flying operations when employing NLW. NLW should not hinder flying operations.

Table V-5. Deny/Neutralize Vehicles, Aircraft, Vessels, and Facilities

Task	***NLW Planning Factors***	***Equipment***	***Employment***	***Notes***
Deny/ neutralize vehicles, aircraft, vessels, facilities	Administration/Personnel • SJA/claims planning • Media plan	Support Equipment • Sensors to track approaching vehicles, aircraft, and vessels • Night vision equipment • Materiel handling equipment • Portable bullhorn	• Employ ground and water sensors, if required • Employ aerial or naval pickets • Use materiel handling equipment to evacuate vessel	• Commander needs to define neutralization mission • Element of surprise remains a critical factor in mission accomplishment • Commander needs to develop IFF procedures
Deny/ neutralize vehicles, aircraft, vessels, facilities	Intelligence • Sensor emplacement if required • Threat ID • IPB • Analyze target area			
Deny/neutralize vehicles, aircraft, vessels, facilities	Operations • Prepare and confirm ROE • Integrate NLW and deadly force • Allocate NLW and designate unit use • Determine scope of mission • Determine access procedures • Determine boarding procedures (vessels and aircraft) • Determine apprehension procedures • Determine decontamination (DECON) procedures	RCA • Marking agent	• Use marking agents to mark a vessel for later interception	

Table V-5. Deny/Neutralize Vehicles, Aircraft, Vessels, and Facilities

Task	***NLW Planning Factors***	***Equipment***	***Employment***	***Notes***
Deny / neutralize vehicles, aircraft, vessels, facilities	Logistics • Distribute NLW munitions • ID special storage requirements • Resupply NLW munitions • Determine vessel/aircraft disposition	Riot Control • Xenon searchlight	• Use RCA to incapacitate crew members • Searchlights can disorient personnel or illuminate target	• Commanders must consider flying operations when employing RCA • Area(s) should be reusable in short order.
Deny / neutralize vehicles, aircraft, vessels, facilities	Civil Affairs • Collect civilian intelligence • Handle displaced civilian personnel	Entanglements • Caltrops • Foams • Speed bumps • Nets • PVAB	• Use caltrops to disable vehicle • Use combination of foam and caltrops to preserve the element of surprise • Channel passive traffic away • Use nets to foul props	
Deny/neutralize vehicles, aircraft, vessels, facilities	Medical • Treating injured civilian personnel			

Table V-6. Seize Personnel

Task	*NLW Planning Factors*	*Equipment*	*Employment*	*Notes*
Seize personnel	Administration/ Personnel • SJA/claims planning • PAO media plan	Support Equipment • Optics	• Employ optics with DM to ID personnel to be seized • Cover personnel employing nonlethal force	• Commander needs to define seizure mission and strictly enforce ROE • Element of surprise is critical for mission accomplishment • Speed is critical • Seizing personnel can be surgical with special operations personnel or conducted within a crowd to seize an agitator
Seize personnel	Operations • Prepare and confirm ROE • Integrate NLW and deadly force • Allocate NLW and designate unit use • Integrate NLW fires and maneuver • Breaching and assault planning • Determine holding area/processing center	RCA • OC dispenser • CS	• Use RCA to incapacitate personnel • Spray into eyes and face for full effect	
Seize personnel	Logistics • Distribute NLW munitions • ID special storage requirements • Resupply NLW munitions • Plan transportation	Riot Control • Xenon searchlight • Flex cuffs • RCA markers	• Searchlights dazzle/disorient individuals • Use flex cuffs for apprehension • Use RCA markers to mark agitators for seizure	

Table V-6. Seize Personnel

Task	*NLW Planning Factors*	*Equipment*	*Employment*	*Notes*
Seize personnel	Civil Affairs • Collect civilian intelligence • Care for injured personnel • Detain civilian personnel			
Seize personnel	Medical • Medical evacuation (MEDEVAC) planning			

Table V-7. Clear Facilities of Personnel

Tasks	*NLW Planning Factors*	*Equipment*	*Employment Considerations*	*Notes*
Clear facilities of personnel	Administration/ Personnel • SJA/claims planning • Chaplain affairs • Casualty notification • PAO media plan	Support Equipment • Riot face shield • Full-length riot shield • Expandable riot baton • Optics • Portable bullhorn	• Distribute to troops in advance • Do not employ riot baton to the head • Employ optic with DM to ID occupant leaders and provide lethal cover fire • Use bullhorn used for communicating with occupants	• Surprise is a critical element for mission accomplishment • Commander must strictly define ROE • Due to close quarter tactics, the force must be instantly ready to utilize deadly force if required

Table V-7. Clear Facilities of Personnel

Tasks	*NLW Planning Factors*	*Equipment*	*Employment Considerations*	*Notes*
Clear facilities of personnel	Intelligence • IPB • Background on facility, access, power plant • Route/map recon • Background info on building occupants • Determine any outside support for building occupants	Kinetic Rounds • 12-gauge bean bag round • 12-gauge rubber bullet • 12-gauge launching cartridge • 40-mm rubber baton • 40-mm wooden baton **NOTE:** USMC no longer uses wooden baton. • 40-mm stinger Stun grenade • Flash bang	• Do not employ kinetic rounds at less than 15 feet due to possible fatal outcome • Targeted personnel at 15 to 40 feet should be center mass • Head shots are not acceptable • Kinetic rounds will not be skip fired • Kinetic rounds effective against selected targets and to disperse individuals • Stun grenade effects can be amplified in buildings • Visibility will be reduced once kinetic rounds are employed • M203 and shotguns are nonlethal complementary. M203 has a low rate of fire, area target. Shotgun has a high rate of fire, point target	• Using RCA greatly reduces visibility • Use of RCA is restricted to certain circumstances; verify that proper authority has been given before use

Table V-7. Clear Facilities of Personnel

Tasks	*NLW Planning Factors*	*Equipment*	*Employment Considerations*	*Notes*
Clear facilities of personnel	Operations • Tactically integrate NLW and lethal fires • FSE plans to include RCA • Allocate NLW munitions • Integrate NLW fires and maneuver • Determine assault and breaching plan • Plan for isolation of facility	RCA • OC dispenser • Team OC dispenser • High-volume dispenser	• Disrupt/disperse crowds • Apply RCA to face and eyes for full effect	• Isolate area to prevent reinforcements • Prepare disposition plan for cleared personnel • Consider use of facility • Prepare security plan for cleared facility
Clear facilities of personnel	Logistics • Distribute NLW supplies • Resupply NLW munitions • Plan for transportation of displaced civilians • Plan facility maintenance or repair	Riot Control • Xenon searchlight • Flex cuffs • RCA markers • Flash bangs	• Searchlights dazzle/disorient individuals • Use flex cuffs for agitators • Use RCA markers to mark agitators for future ID • Flash bangs to dazzle/disorient individuals	
Clear facilities of personnel	Civil Affairs • Collect civilian intelligence • Care for injured civilians • Coordinate plans for controlling civilians			
Clear facilities of personnel	Medical • Hospital planning • Medic planning for tactical units • Procurement of medical supplies			

4. Staff Planning/Responsibilities

a. Administration. A staff composed of the G-1/S-1, PAO, SJA, and others as needed, is responsible for advising the commander on NLW employment and ROE. Additionally, administrative personnel are responsible for processing all civilian detainees and claims that arise from NLW employment. As a minimum, the staff must consider the following:

(1) SJA/claims planning.

(2) Chaplain.

(3) Casualty notification.

(4) Detainee disposition plan.

(5) PAO/media planning.

(6) Information dissemination.

b. Intelligence. The G-2/S-2 office is responsible for conducting the IPB process and providing information on the adversary's culture, political stability, and level of dedication to the cause, motivation, leadership, capabilities, equipment, and tactics. The G-2/S-2 should locate native language speakers to assist the interrogator-translator team.

(1) Background on country and culture.

(2) Route/map reconnaissance.

(3) Place sensors for advance notification.

(4) Information on leaders.

(5) LP/OP.

(6) Determine reason for the crowd.

(7) Identify threat.

(8) Analyze traffic.

c. Operations. The G-3/S-3 office is responsible for unit training, nonlethal certification of units, mission analysis, maneuver plans, and CA planning when required.

(1) Tactically integrate NLW and lethal fires.

(2) Fire support plans to include using RCA.

(3) Allocate NLW munitions.

(4) Plan engineer barrier.

d. Logistics. The G-4/S-4 office is responsible for logistics considerations, ship-to-shore movement of weapons and ammunition, storage and maintenance of equipment, resupply, transportation ashore, and other support requirements inherent in using NLW.

(1) Distribute NLW supplies.

(2) Security of rear area facilities.

(3) Resupply NLW munitions.

(4) Special maintenance requirements for NLW equipment and munitions.

e. Communications Officer. The G-6/S-6 staff is responsible for installing, operating, and maintaining communications from units confronting belligerents to the tactical operation center (TOC) or combat operations center (COC). The staff must plan for the effects NLW will have on communications to include proper frequencies.

f. PAO. The PAO provides liaison with outside media and participates in the staff planning process. PAO is also responsible for ensuring the NLW media plan and news releases consider the perceptions of the native media and belligerents or adversaries. PA personnel must receive training on NLW technologies, employment and capabilities, acute and long-term effects, impact on the environment, and legal approval procedures (at a minimum) so they can respond to media and public inquiries.

g. Medical. Medical personnel must be aware of the effects of NLW. Medical personnel should also be familiar with and able to provide decontamination of RCA as well as administer antidotes in cases of adverse reactions to riot control and other agents.

(1) MEDEVAC planning.

(2) Procuring medical supplies.

(3) Hospital planning.

(4) Medical planning for tactical units.

h. Chaplain. The commander should consider chaplain support for counseling and/or religious services for military and civilian personnel. Additionally, chaplains should provide information regarding the religious and cultural background of the local populace.

i. CA. The CA office is responsible for planning and coordinating all aspects of CA activities and providing assistance in gathering intelligence from civilian sources.

(1) Handle displaced civilian personnel.

(2) Coordinate plans for control of civilians.

j. PSYOP. Tactical PSYOP forces serve as an ideal tool for the commander to resolve issues of armed resistance and local interference with military operations while encouraging dialogue and cooperation with noncombatants. This force provides dissemination of PSYOP products (loudspeaker messages, leaflets, posters, etc.) and conducts face-to-face communications with target audience. PSYOP forces provide the supported commanders with nonlethal leverage by influencing foreign populations and expressing information subjectively to influence attitudes and behavior, thereby obtaining compliance or noninterference. They provide public information to foreign populations to support humanitarian activities, ease suffering, and restore or maintain civil order. They serve as the supported commander's voice to foreign populations to convey intent, countering enemy propaganda, misinformation, and opposing information to correctly portray friendly intent and actions. Finally, PSYOP forces deny others the ability to polarize public opinion and political will against the United States and its allies within an AO.

5. Tactics

The commander should include the following considerations in preparing for unit tactical operations involving NLW:

a. Avoid reconfiguring small-unit organization, except for special configurations such as—

(1) Special teams.

(2) Attachments.

(3) Crowd control formations.

(4) Security organizations.

b. Standardize postures in unit tactical SOP. Take the following steps to add a higher volume of specialty impact munitions and equipment within squads, platoons, or units:

(1) Avoid arming troops involved in apprehending designated personnel with long weapons (i.e., shotguns, rifles) because they can be easily grabbed/pulled away by individuals.

(2) Add weapons systems (shotguns) that provide a greater nonlethal capability. When loaded with nonlethal munitions, the longer reloading time of the M203 grenade launcher is protected by coverage from the high rate of fire from shotgun.

(3) Add nonlethal munitions to existing organic weapons systems (M203 grenade launcher).

(4) Avoid arming troops in the front lines of crowd control formations with long weapons.

(5) Focus riot-control capabilities in specialized reaction forces.

(6) Establish base line NLW postures.

(7) Establish command relations for special units.

(8) Maximize distance and barriers.

(9) Create NLW range cards for static positions.

(10) Provide the force with available OC/RCA consistent with ROE.

(11) Always mutually support forces utilizing NLW with lethal means.

(12) Consider environmental conditions (heat/cold/wind) and their effect on the performance of nonlethal munitions. Proper storage, periodic inspections, and rotation of NLW ammunition are critical to maintaining the effectiveness and viability of nonlethal munitions.

c. Unit Weapons, Unit Configurations, and Ammunition Considerations.

(1) Individual. Designate individuals as nonlethal shooters. Nonlethal shooters should also carry lethal munitions for use as required. Lethal rounds must be marked and carried in a manner to prevent confusing nonlethal rounds with lethal rounds.

(2) Squad. Squad leaders should dictate who carries nonlethal grenades and maintain their weapons loaded with lethal ammunition. Ideally, the squad should not change its task organization to accommodate nonlethal equipment, but should designate nonlethal shooters.

(3) Patrols. Commanders should not plan a nonlethal patrol, but should plan a combat/security patrol with a nonlethal capability when mission dictates.

(4) Static Positions. Individuals in static positions should have their weapons loaded with lethal munitions (mission, enemy, terrain and weather, troops and support available–time available and civil considerations dependent) (METT-T/METT-TC). Additionally, they should carry nonlethal munitions. Both shotgun and M203 work well from static positions and should complement each other in a nonlethal mode.

(5) Recovery Teams. Recovery teams have the primary responsibility of retrieving and controlling designated personnel from the crowd; the commander should establish a recovery team before deploying. The force should train team members in unarmed self-defense, open-hand control, and hand cuffing/flex cuffing. It should equip recovery teams with personal protection gear, 9-mm weapon, and flex cuffs. Due to the physical nature of recovery teams, it is important not to arm team members with long rifles. If security for the recovery team is an issue, the commander should attach a security element.

(6) Crowd Control Formations. The unit should establish crowd control formations within a minimum response time. Because of the physical nature of crowd control, individuals in these formations, if possible, should not carry long rifles. These formations should have inherent nonlethal capability. The unit must provide lethal coverage for the entire formation.

d. Individual Elements of a Formation. These elements are: base element, support element, and command element. The formation needs to provide versatility. Restrictions must never be set on numbers, equipment, or changes to the standard formations.

(1) Base Element. Role of the base element (front line):

(a) Base element serves as the front line or the base/foundation of the formation (can be divided into two or three sections or ranks).

(b) Front line consists of the shield holders.

(c) Angle the bottom of the shield to channel flammable liquids away from the body.

(d) Distance or interval between the shields differs based upon the range to the crowd. The closer the crowd, the tighter the interval between the shields.

(e) Provide frontal security.

(f) Hold a solid line.

(g) Provide intimidation factor.

(h) Fire team leaders are in the line holding shields.

(i) Secondary (tertiary) line(s) can consist of the baton holders and/or nonlethal gunners.

(j) Baton holders will—

- Strike with batons.
- Spray OC or disperse smoke.
- Serve as spotters for the NLW gunners.
- Control and detain subjects pulled from the crowd.

(k) Nonlethal gunners will—

- Consist of team leader with M203/M16 (provides nonlethal cover with M203 40-mm munitions and lethal cover with M16).
- Position themselves between shields working with baton holders.
- Serve as immediate control for the team.
- Maintain continuous muzzle awareness of the weapon.
- Provide immediate support to the shields.

(2) Support Element. Role and employment of the support element:

(a) Provide the means to employ lethal and nonlethal capabilities.

(b) Can consist of two ranks: M16/M203 (2nd rank) Mossberg 12-gauge shotguns (3rd rank).

(c) Provide security for search and recovery team.

(d) Consists of munitions teams and/or recovery teams.

(e) Munitions teams provide the means to employ lethal and nonlethal capabilities consisting of M16/M203 and Mossberg 12-gauge shotguns.

(f) Recovery teams provide ability to detain and search rioters downed by nonlethal munitions. Recovery teams will—

- Extract vehicles or personnel from the crowd who are in immediate danger.
- Detain and escort downed rioters to rear of the formation.
- Restrain and search detained individuals.
- Communicate with the base element.
- Spray or employ RCA as needed.
- Fill in the gaps in the front line as needed.
- Insert into a confined or congested area too large for the formation.
- Use a soft base line that capitalizes on the munitions fired to deploy forward of the shields in order to drive the crowd back, retrieve personnel/gear, and apprehend key riot personnel. Employed in pairs (1 x shield man/1 x shotgun).

NOTE: Refer to Marine Corps Security Force (MCSF) formation examples.

---

(3) Command Element. Role of the command element is to provide command and signal to the formation and consists of the platoon leader/flight commander/ platoon sergeant/flight sergeant, and squad leaders.

(a) Platoon Leader/Flight Commander, Platoon Sergeant/Flight Sergeant—

- Pass verbal commands.
- Maintain order within formations.
- Gauge and evaluate use of munitions and resupply considerations.

(b) Squad Leaders—

- Effectively control the squad.
- Pass information up the chain.
- Employ OC as directed.

(c) Designated Marksman (DM). During a nonlethal engagement, using DM provides confidence and safety to those facing a riot. The DM in an overwatch position scans the crowd, identifies threats and designated personnel for recovery, and fires lethal rounds if a lethal threat is presented. Additionally, DM are ideally suited for flank security and counter sniper operations.

6. Employment Objectives

a. Innovative employment methods of NLW and ingenuity take on critical significance. NLW are employed in concert with other nonlethal and lethal systems to create a synergistic effect.

b. The measures for countering many nonlethal options are usually apparent, quickly learned, and readily available. Because they are not intended to kill, nonlethal options "teach" an adversary what to avoid in the future. Small unit commanders must prepare to keep one step ahead of motivated belligerents.

c. Many nonlethal munitions have both maximum effective and minimum safe ranges. Individuals short of the minimum safe range may suffer severe injuries or death while the effects of most nonlethal munitions are greatly decreased at longer ranges. To be effective, an adversary must be engaged within the optimum range of the munition. Figure V-1 illustrates the ranges for a variety of nonlethal munitions and systems.

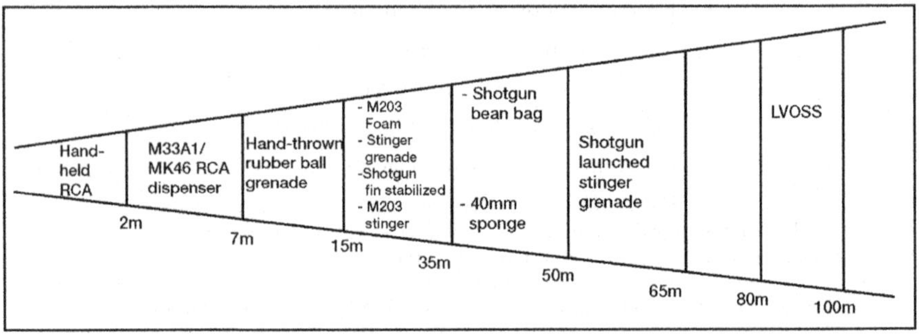

Figure V-1. Nonlethal Weapons Ranges

7. Rules of Engagement

a. U.S. forces follow directives issued by competent military authority that delineate the circumstances and limitations under which deadly force will be initiated against other forces. Intrinsic to the effective use of nonlethal options is the authorization to employ them. ROE are the primary means by which the chain of command can provide guidance to deployed forces in peacetime for handling crises and, in wartime, to help control the level of hostilities.

b. The commander should distinguish ROE from tactical control measures, threat conditions, and arming orders/weapons conditions. Although arming orders are sometimes listed on the same quick reference card as ROE for easy reference, the commander should distinguish ROE from other policies and directives developed, such as weapons confiscation rules.

c. The commander should delegate the decision to use NLW against an adversary to the lowest possible level, preferably to the platoon, flight, or squad level. However, this requires that all personnel, not just leaders, have a clear understanding of ROE and his intent.

d. Using lethal force, employed under the standing ROE, will never be denied. At no time will forces be deployed without the ability to defend themselves against a lethal threat nor will they forego normal training, arming, and equipping for combat. Nonlethal options are a complement to, not a replacement for, lethal force and seek to expand a proactive response across the range of military operations.

e. The following figures are examples of service-specific formations. These examples are not all-inclusive, and the tactical situation will always dictate actual unit employment.

FM 19-15, *Civil Disturbance*; MCWP 3-11.1, *Rifle Company;* and MCWP 3-11.2, *Marine Rifle Squad* provide extensive detail about small-unit formations, their use, and other considerations. Figures V-2 to V-7 are from I and II Marine Expeditionary Force (MEF) Special Operations Training Groups (SOTG) and represent a notional infantry platoon. Figures V-8 to V-13 represent a notional U.S. Army infantry platoon. Figures V-14 to V-21 are additional platoon formations submitted by Marine Corps Security Force Training Company.

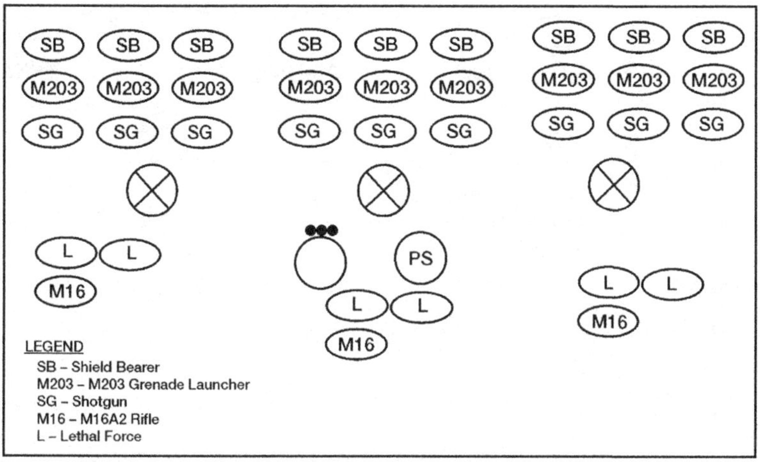

Figure V-2. Platoon in Depth

NOTE: The platoon is the smallest unit capable of independent maneuver. The platoon in depth formation shown above provides the following benefits:

- In-depth support with three ranks.
- Complementary support with the shotguns covering the slower reload time for the M203.
- Additional personnel from each squad to form the recovery teams.
- Each squad is capable of moving as an independent formation forward of the platoon line, if situation dictates.
- Other tactical formations are applicable—echelon, wedge, column.

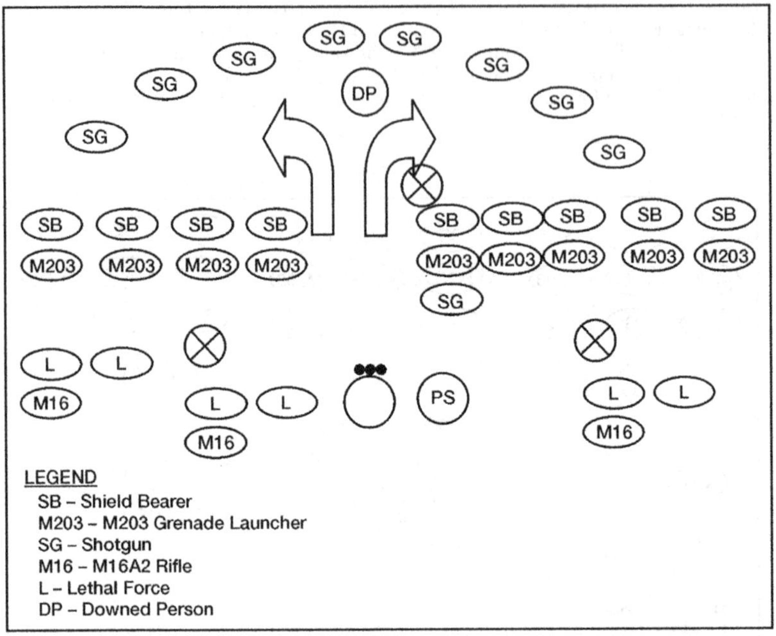

Figure V-3. Platoon Recovery Team Deployment—Security Element

---

NOTE: Security element deploys around downed person.

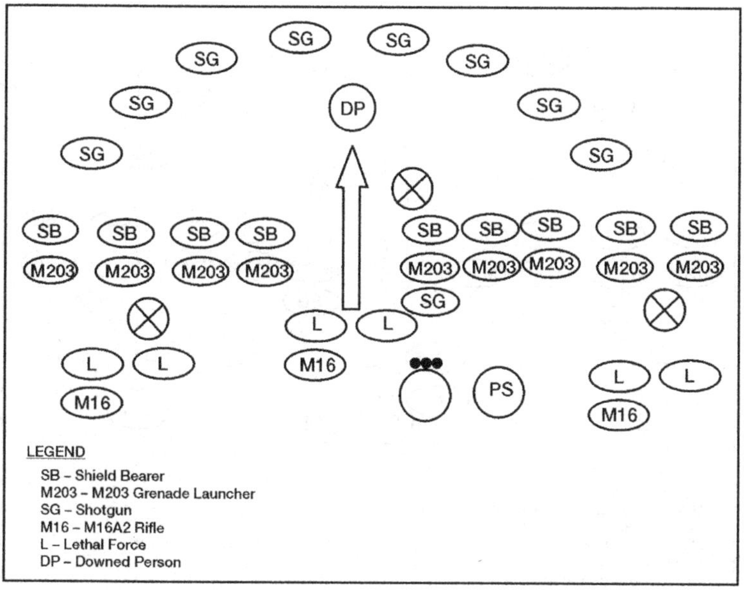

Figure V-4. Platoon Recovery Team Deployment—Recovery Team

NOTE: Recovery team deploys to detain, search, and recover downed person.

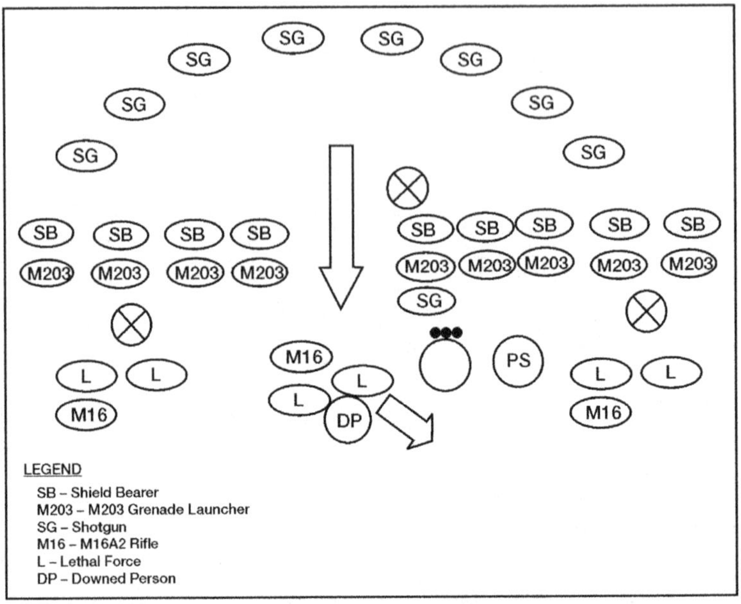

Figure V-5. Platoon Recovery Team Deployment—Recovery Team Redeploy

---

NOTE: Recovery team redeploys with detained person.

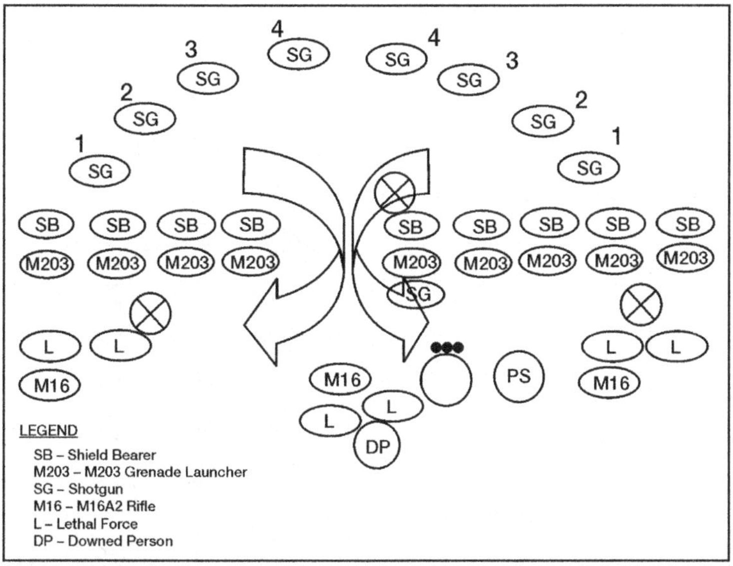

Figure V-6. Platoon Recovery Team Deployment—Security Element Redeploy

---

NOTE: Security element redeploys back to 3rd rank "by the numbers." Team leader calls "4s in," then "3s in," etc. The last person in calls "last man in" and formation closes the gap.

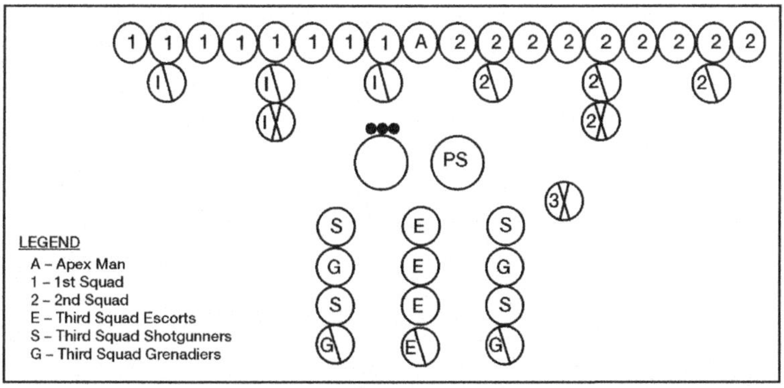

Figure V-7. Platoon On Line—Munitions Squad in Close Support to Cover Flanks

---

NOTE: Platoon in close support. NLW and lethal weapons are sited to cover the flanks of the formation, as well as support the front line.

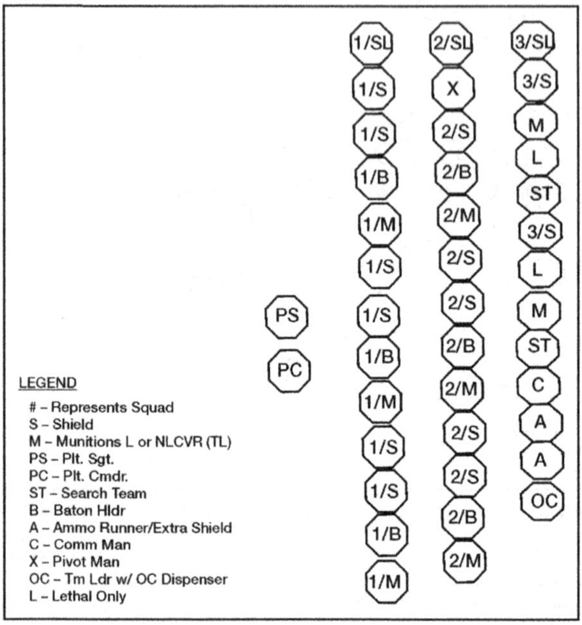

Figure V-8. Platoon Column

NOTE: Platoon in column formation. For administrative movement only.

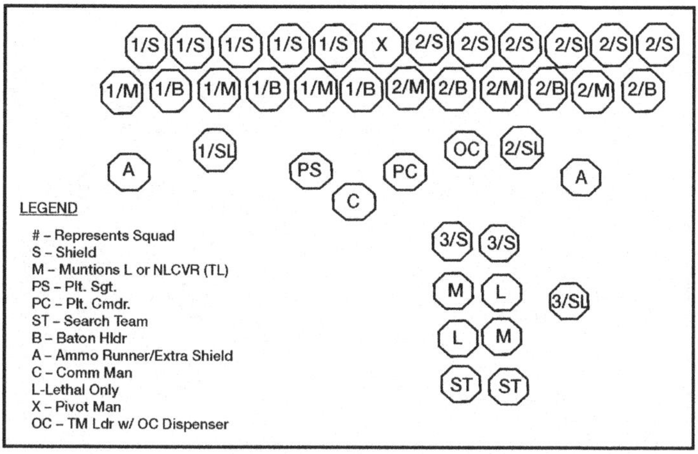

Figure V-9. Platoon On Line

NOTE: Platoon on line—basic formation for the employment of NLW.

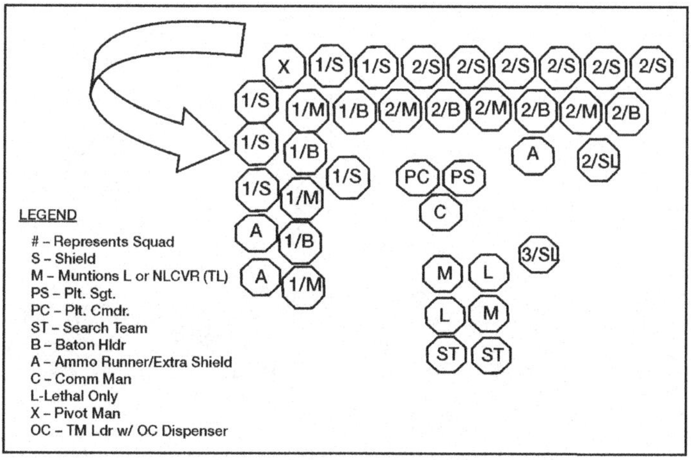

Figure V-10. Platoon Lateral Support, Left (Right)

NOTE: Platoon in lateral support to protect flanks.

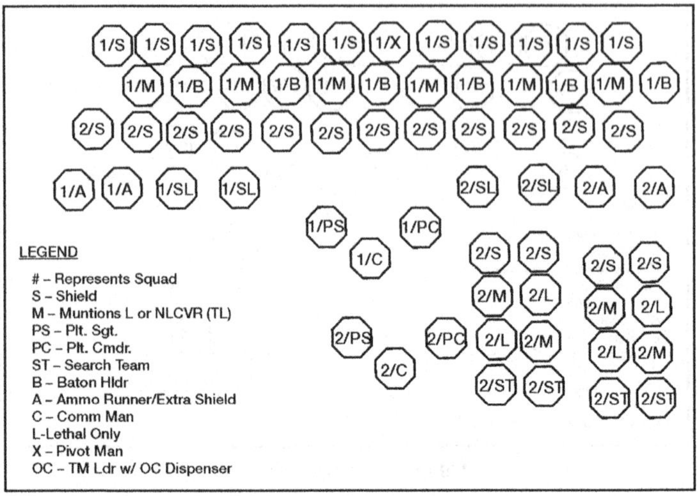

Figure V-11. Two Platoons in Close Support

NOTE: Two platoons in close support—tied in for mutual advantage.

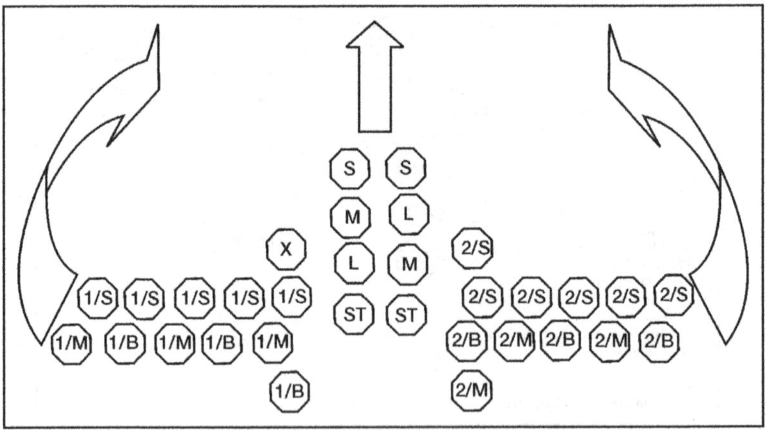

Figure V-12. Recovery Team Deployment

NOTE: Recovery team deploys from formation—remainder of platoon can deploy around the recovery team from either flank.

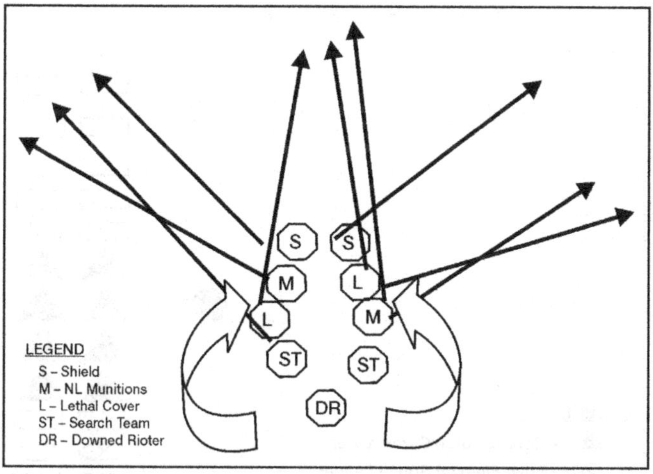

Figure V-13. Recovery Team Actions and Security Posture

NOTE: Recovery team forms security perimeter around downed rioter, conducts search and detain actions. Once actions are complete, recovery team redeploys.

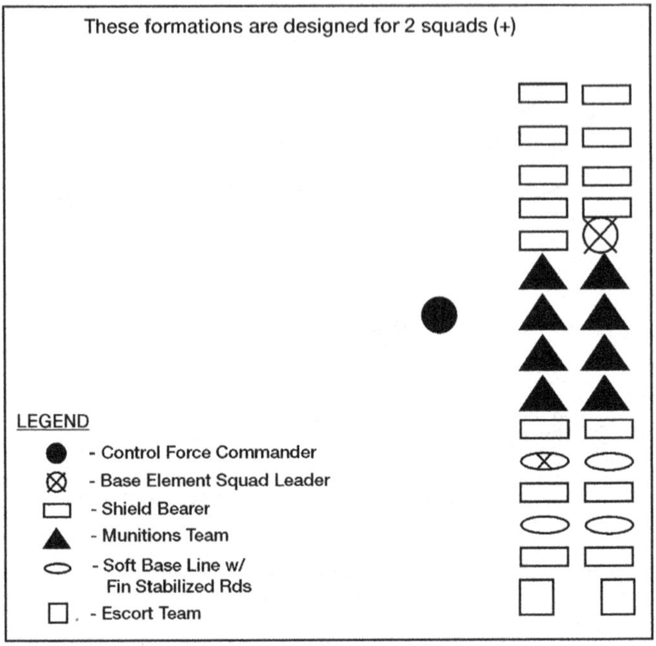

Figure V-14. Platoon in Column (USMC)

NOTE: These formations are designed for 2 squads (+). This allows the platoon to move rapidly when there is no chance of being engaged with the crowd.

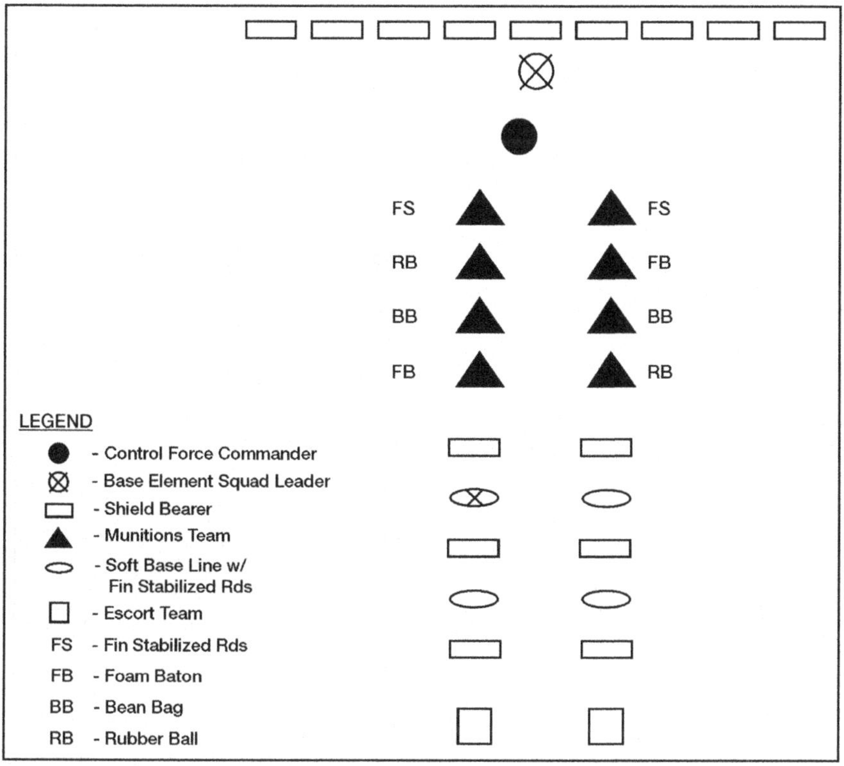

Figure V-15. Platoon in General Support (USMC)

NOTE: The control force commander moves into this formation approximately 60 to 70 meters from the crowd, as the crowd could start to engage the control force with thrown objects.

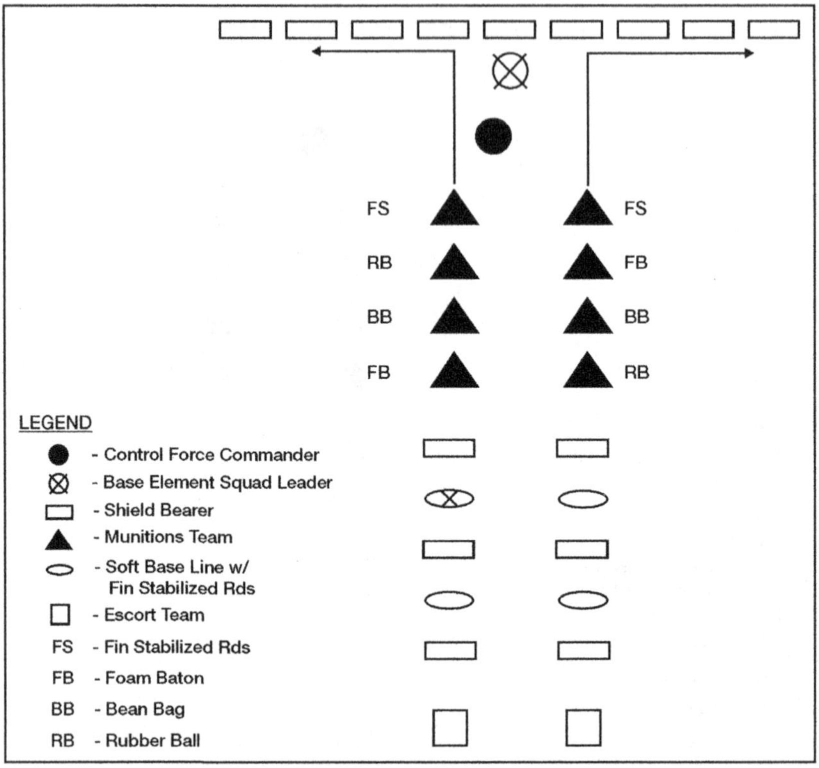

Figure V-16. Platoon in Close Support (Start) (USMC)

NOTE: The control force commander moves into this formation when he feels ready to engage the crowd with nonlethal munitions.

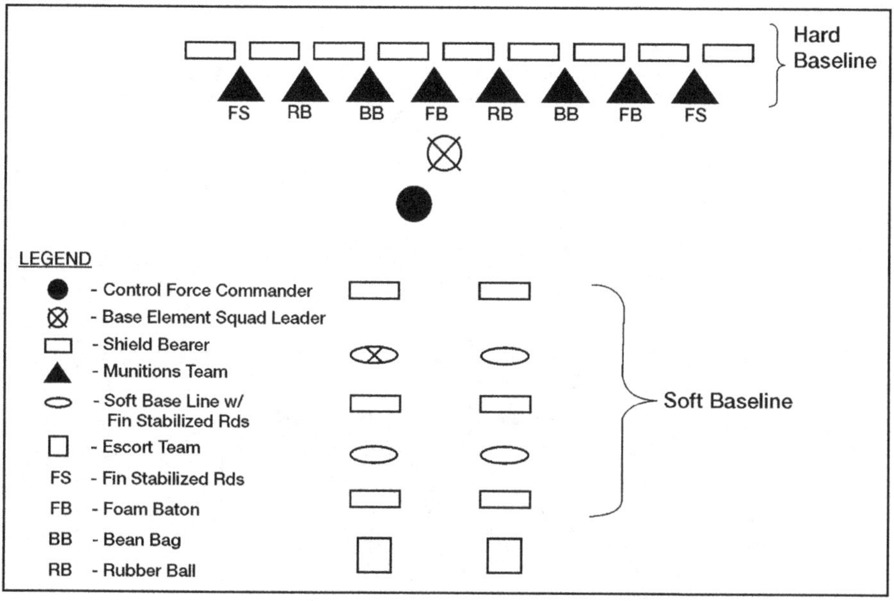

Figure V-17. Platoon in Close Support (Finish)

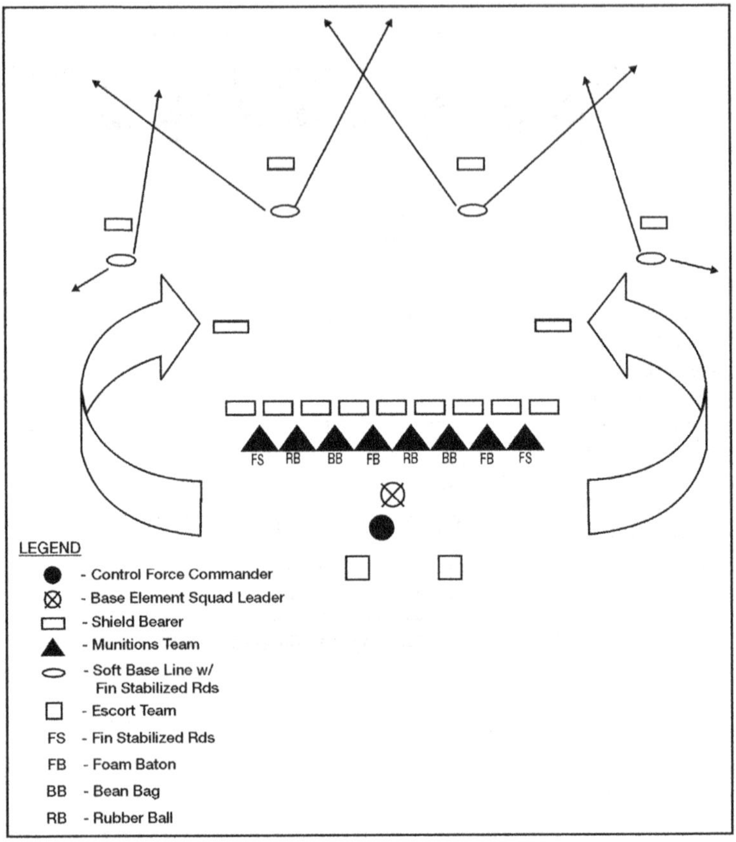

Figure V-18. Deploying Soft Baseline

NOTE: The soft baseline affords the control force commander much flexibility. The primary mission of this formation is to disperse the crowd back using speed and extreme violence of action. By deploying the soft baseline, the commander is able to maneuver the hard baseline where needed (around the barrier plan or obstacles.) The soft baseline is deployed by a predetermined signal (bullhorn siren, whistle blasts, vehicle horn) and will be retrieved using the same signal. This signal conditions the crowd to expect violence of action and causes them to disperse. The soft baseline should never be deployed into a fresh crowd; the crowd needs to be distracted by nonlethal munitions. The formula found to work is distract with stingball grenades, followed by a volley of shotguns and M203 munitions, then send the soft baseline out to disperse or move the crowd back. (Sting them, shoot them, soften them up.)

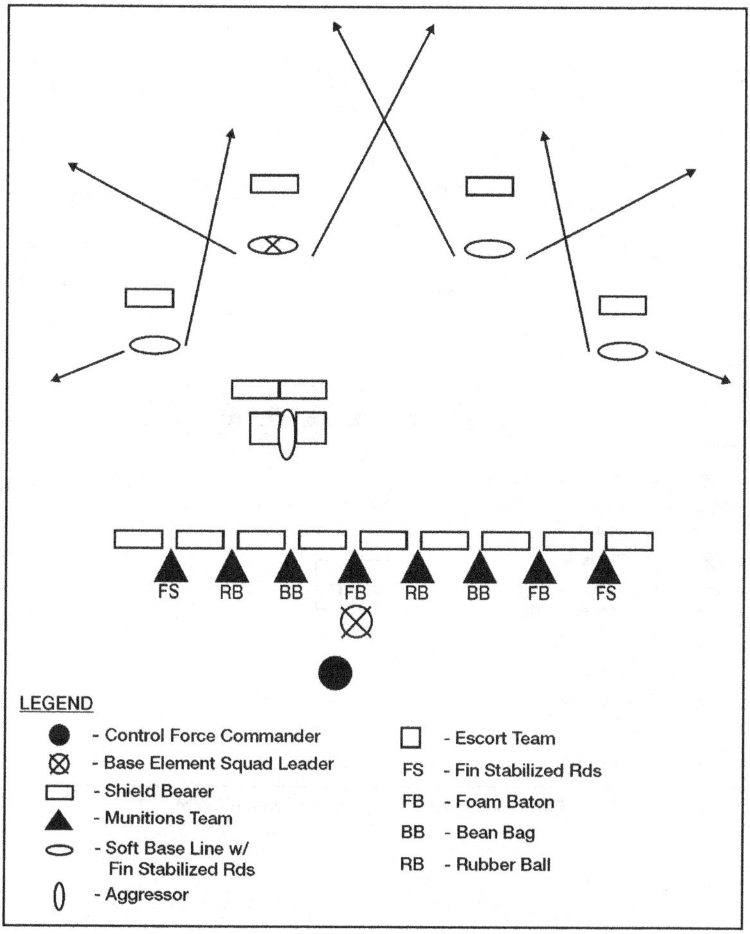

Figure V-19. Deploying Escort Team (Start)

NOTE: When the control force commander identifies the aggressor to retrieve, the command is, "Escort team left/right," then the signal. The signal is the same to deploy the soft baseline. As the soft baseline deploys, the two shields without gunners provide protection for the escort team. The shields remain with the escort team until the soft baseline is called back in, at which time the shields go back to the soft baseline.

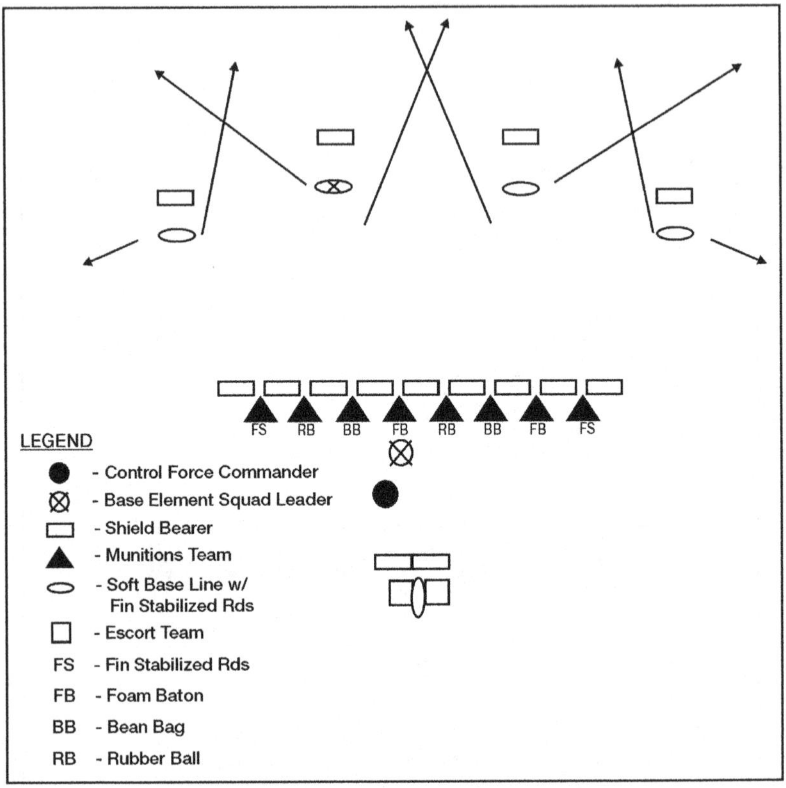

Figure V-20. Deploying Escort Team (Finish)

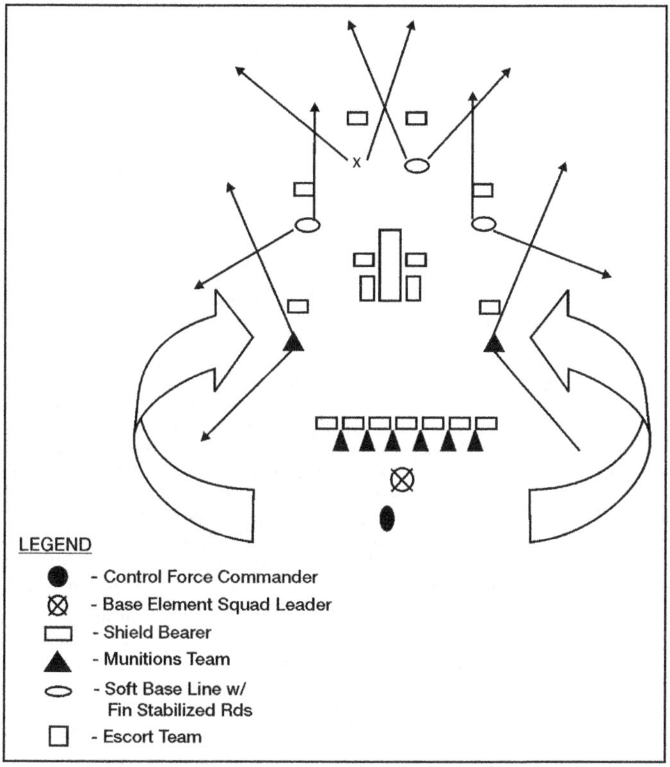

Figure V-21. Recovery of Downed Personnel or Stranded Vehicle (Big Johnson)

NOTE: In the event the crowd attacks personnel or a vehicle, the control force commander can call for a "Big Johnson." In a Big Johnson, the soft baseline deploys as usual except now the shield and shotgunners on the flank of the hard baseline also deploy creating a 190-degree perimeter around the personnel or vehicle. This maneuver must be executed quickly and with extreme violence of action because the crowd already has its hands on individuals in the force. The escort team deploys with the Big Johnson in order to escort or carry injured individuals. The shields without gunners provide protection for the escort team.

Chapter VI

LESSONS LEARNED

1. Background

Although lessons learned from employing NLW are still being studied and impacts analyzed, it is beneficial to look at the preliminary lessons gained from the NLW experience. These lessons are far from conclusive, because U.S. forces have yet to see the first large-scale employment of NLW. However, U.S. forces can gain tremendous insight from the experience of others and use what has been learned to guide future NLW employment.

2. Lessons Learned

a. Force Continuum. Using nonlethal options needs to be based on the principle of a graduated and proportional response, using the decisive force necessary to respond to the threat.

b. Attempts to legislate and regulate every aspect of NLW employment can prove disadvantageous to the on-scene commander. During the evacuation of U.N. forces in Somalia (OPERATION UNITED SHIELD, 1995), the ROE made an arbitrary and unnecessary distinction between situations justifying deadly force and those that justified only lesser means. Crowd control items were restricted to situations requiring deadly force. Many soldiers, airmen, and Marines asked, "why resort to nonlethal means at all if I have to wait until deadly force is authorized before I can shoot them with a beanbag?" Additionally, the use of sticky foam was limited to deadly force situations because of the perception that adversaries might suffocate from its application. This stipulation effectively eliminated sticky foam as a nonlethal option to the forces in Somalia.

c. With ROE that take into account the force continuum and encourage graduated response to nonlethal threats, the commander can employ NLW to enhance his flexibility, deterrence ability, and reaction to uncertain situations. To accomplish the mission with minimum loss of life and property, commanders need to have ROE that do not restrict them from properly employing NLW.

Lessons Learned from OPERATION JOINT GUARDIAN 1999 (Battalion Landing Team (BLT) 3/8 26 Marine Expeditionary Unit (MEU) Peace Operations in Kosovo:

ROE significantly impaired the employment of NLW. For example, OC was not authorized because it was considered a chemical agent. Nonlethal direct fire munitions were not authorized during the initial occupation of the province. This significantly hampered the individual Marine's ability to neutralize a threat where deadly force would not have been justified but action was required. Overly restrictive ROE thus negated the flexibility of response that NLW are supposed to provide. The intent of deploying NLW is to provide the operator the capability and flexibility to neutralize a threat without having to apply deadly force. Thus, the ROE for NLW should ideally be less restrictive, not the same or more restrictive, than those for lethal force.

It is often enticing for authors of ROE involving the use of NLW to base those ROE on the principle of minimum force necessary. This may be appropriate for base and station or homeland security ROE; however, for expeditionary operations the decision to mandate minimum force deserves critical consideration. Commanders in military operations have traditionally gained advantage over their enemy not only by the actual combat power that they bring to the fight, but also from that enemy's perceptions and beliefs about the intensity and resolve with which that combat power will be applied. When ROE are written stating that the confrontations will be addressed using the minimum force necessary, this advantage is lost. ROE developed for expeditionary nonlethal operations should be based on the concept of decisive and compelling force, rather than the minimum force. This will maintain an adversary's hesitancy to harass and endanger U.S. forces even after that adversary has tested or probed the response of those U.S. forces. Further, this will allow synergy to be gained through the complimentary employment of lethal and NLW by providing more response options to the small unit leader.

3. Employment

a. Through the use of modeling, simulation, and field exercises, new concepts for employing NLW are continually being developed. The force should train with NLW to the same degree as for lethal weapons. In this way, units become confident when using nonlethal options in chaotic and uncertain situations. The commander should push the decision to employ NLW to the lowest possible tactical level. Commanders who intend to use NLW in achieving military objectives must recognize the fluidity of the situations in which they are likely to be employed. The decisionmaking cycle is short, stressful, and the level of violence can move in both directions along the force continuum. In quick succession, a small unit leader might employ nonlethal munitions, decide to employ deadly force to an immediate specific threat that arises, then return to using nonlethal means in response to unarmed hostile elements.

b. A mixture of nonlethal "tools" has been determined to be advantageous in an operational scenario. As the situation develops, commanders can apply different nonlethal responses with respect to range/distance capability and severity of effect. For example, to disperse a crowd, area fire weapons would precede use of close-in point target nonlethal munitions.

c. An important point to remember is that NLW are just one element of a graduated response available to the force commander. Other factors make up a large contribution to the success of U.S. forces, including barrier plans and PSYOP. An example of this operation would be using barriers such as roadblocks or concertina wire, warning commands to inform a mob of potential responses based on their actions, and only then taking actions through escalated employment of NLW. These continued responses would involve smoke, RCA, and blunt munitions. A specific verbal warning should precede every escalation of nonlethal means.

d. Whenever NLW are employed, a separate lethal capability must always be in place. Troops must always be prepared to protect themselves as required by lethal means, because it is impossible to accurately predict the response of an individual or mob to NLW employment.

Lessons Learned from OPERATION ESSENTIAL HARVEST 2001 (U.S. Embassy Reinforcement operations conducted at the U.S. Embassy Skopje Macedonia by 5th Platoon, 2nd Fleet Antiterrorism Security Team [FAST] USMC):

The anticipated threats were violent demonstrations directed against U.S. facilities in the AO. In this case, USMC forces tasked to reinforce the embassy were equipped with a compliment of NLW. ROE took into account the capabilities and limitations of the NLW. The ROE and the NLW systems on the ground were coordinated and thus compatible. In addition, it was found that integrated employment of different nonlethal munitions greatly enhanced that force's ability to suppress the threats encountered (large rioting mobs). Random security patrols reinforced with military working dogs proved very effective. Reinforcement with multiple M203s provided USMC units with both a lethal and nonlethal capability in the same weapons system. The M203 was particularly useful as it gave the individual Marine the means to quickly transition between a lethal (5.56-mm) and nonlethal (40-mm nonlethal) capability. A basic understanding of the capabilities and limitations of nonlethal munitions enabled a relatively small force (40-50 Marines) to counter a sizable threat many times its size (500-1000 rioters). Although NLW were not used, the capability and preparation of the units provided the adequate deterrence that prevented the situation form escalating.

4. Publicity and Media

a. The commander can use the media to his advantage, as they will inform the local population of the military's capabilities and responses available. During OPERATION UNITED SHIELD in Somalia, the local media was informed of the Marines' nonlethal options and "new technologies" in order to discourage hostile actions. The media was encouraged to report this to the population. Intelligence had indicated the Somalis were planning to mix civilians with armed clansmen, creating hostile mobs. In fact, feuding warlords had previously successfully demonstrated this strategy on several occasions. Lack of deadly mob incidents implies Somalis were not anxious to test this new technology. Evidently, publicity had directly contributed to mission accomplishment.

b. The public dissemination of the presence of NLW capabilities must strike a balance between information sufficient to discourage hostile actions and information that enables the local population to learn too much of the capabilities of NLW. Too much information allows adversaries to develop countermeasures or avoid the NLW effects. Commanders who employ NLW systems must involve the PAO from the beginning of planning. The PAO must be familiar with the objectives of the mission, commander's intent, ROE, NLW and lethal weapons employed, and be prepared to advise the commander on the best methods of dealing with outside media. The commander should fully integrate the PA plan with PSYOP objectives and the concept of operations. Additionally, he must work the media to ensure the "sensationalism" of such a new technology does not create unrealistic expectations where he can be unfairly judged or constrained.

5. Capabilities and Impact on Mission Accomplishment

a. The commander should understand the unique capabilities of NLW and use them to accomplish the mission. He should use NLW in conjunction with the planned employment of the force and within established ROE to minimize loss of life and property while accomplishing the mission. The force requires early hands-on training to recognize the capabilities and limitations of NLW. Appendix D has sample training plans developed to meet these objectives.

b. The force commander must fully integrate planning for using nonlethal capabilities as quickly as possible in the staff planning cycle. In order to have a full range of alternatives for use against unarmed but hostile elements, the force needs to deliver nonlethal systems, complete necessary ROE and legal review, and make sufficient quantities of each NLW readily accessible to the force. The choice of nonlethal capabilities should be based on several factors and the choice can be made after addressing these questions:

(1) First, are the NLW systems available and can they be delivered on time?

(2) Second, are they compatible with, and do they complement existing weapons and training?

(3) Third, are there clear and unambiguous guidelines and ROE for employment?

(4) Finally, can the NLW be effectively employed to save lives and contribute to mission accomplishment?

c. Leaders need to anticipate local civilians developing effective countermeasures to their NLW and tactics. Therefore, prior to beginning their mission, leaders must determine and rehearse actions to take when nonlethal munitions are effectively countered.

d. Using nonlethal and lethal force are not mutually exclusive. In many instances, the commander can gain synergistic effects by using a combination of both. The following are some examples of this:

(1) Utilizing NLW to separate agitators from noncombatants with simultaneous employment of lethal force against hostiles or combatants where deadly force is justified.

(2) Utilizing NLW when making entry into structures or rooms to gain time in order to discern temper and intent of occupants prior to applying deadly force.

6. Knowing Your Adversary

a. Although commanders desire thorough intelligence of any potential adversary, such a goal may be difficult to reach in a NLW employment environment. During OPERATION UNITED SHIELD in Somalia and OPERATION UPHOLD DEMOCRACY (1995) in Haiti, a study of the local population's culture, history, and customs enabled commanders to anticipate and control many situations before they got out of hand. On the other hand, during OPERATION UPHOLD DEMOCRACY, poor intelligence and a lack of understanding between the two cultures led to the Haitians and Cubans being lodged together in Guantanamo Bay. This situation led to riots between factions who had a deep dislike for each other. By understanding a particular crowd's motivation, friendly forces may be better able to take necessary steps to prevent escalation resulting in the use of deadly force. Early training in crowd dynamics and riot control will enhance the effective use of NLW.

Lessons Learned from U.S. Peace Operations, 739th MP Battalion (Bn), January 2001

An Army MP platoon received a mission to clear a rioting crowd of 300 civilian locals from a main supply route. The platoon placed vehicle-mounted machine guns in overwatch positions to allow quick transition from NLW to conventional rounds. 12-gauge shotgun teams armed with fin stabilized rounds carried M9 pistols for lethal backup. Two volleys of shotgun and M203 nonlethal rounds successfully dispersed the crowd. The platoon had utilized INIWIC graduates to conduct preparatory training on using NLW. The NLW were extremely effective in pushing back the crowd and causing it to disperse. The 40-mm rubber ball round was moderately effective at close range, but very inaccurate at longer ranges. Shotguns were more effective than the M203 due to their high rate of fire and quick-loading capabilities.

b. Specific lessons learned during this operation are as follows:

(1) Placing interpreters is critical. The solution in this incident was to place interpreters approximately 25 meters behind formation so they could deploy forward as needed. Interpreters were often needed to communicate with crowds and with adjacent forces and coalition units that were organic to Kosovo forces (KFOR) that also responded to the incident.

(2) Leaders as well as soldiers should prepare to speak to the media. The command should have a PAO representative on hand to deal with the media.

(3) The commander should thoroughly brief the transition to lethal force from NLW.

(4) U.S. commanders in Kosovo engaged in the decisionmaking process had disparate levels of knowledge of nonlethal capabilities, constraints, and operational considerations. Therefore, leader training and orientation requires more emphasis.

(5) ROE were vague regarding the use of NLW. In many operations, NLW use is delegated to the on-site commander. Delegating to the lowest possible level is the preferred method.

(6) A unit needs a mixture of nonlethal rounds based on the size of crowd, composition of the crowd, available weapons systems, and synergistic effect of selected nonlethal combinations.

(7) When a RRF/QRF arrives on station with NLW to support a force equipped only with lethal means, the commander should address the following command and control concerns:

(a) Who makes the decision to employ NLW?

(b) Is the reinforced commander familiar with NLW?

(c) Who controls the reinforced unit?

(d) How does this relationship change as the situation develops?

7. Summary

a. The wide range of military operations in which U.S. forces are involved has demanded the introduction of NLW into the arsenal of weaponry available to commanders. NLW provide a graduated and proportional response to threats posed by a variety of adversaries. The ROE must be consistent with the rapidly changing situation of today's chaotic and uncertain military operations. This reality demands a high level of training in properly employing NLW.

b. Effective use of NLW involves a synergy between PSYOP activities, intelligence gathering, and professional media/PA plan. NLW effectiveness is enhanced through integration of these activities to reduce the chance adversaries will be able to develop effective countermeasures. NLW, when combined with lethal weapons, provide the commander with a force response across a wide range of military operations.

Appendix A

DOD Nonlethal Weapons and Equipment Item Description

The following is a cumulative list of NLW and equipment currently fielded within the DOD. Each service has various components of the following list in its inventory. A list of the service's items is located in Appendix E.

Figure A-1. Riot Face Shield

1. Riot Face Shield: The riot face shield is a non-ballistic face shield designed to protect the service member's face from objects thrown or attacks from non-ballistic weapons. (Figure A-1)

No picture available.

2. Ballistic Face Shield: The ballistic face shield provides the service member with face protection from a wide range of threats (i.e. debris, liquids, handthrown objects). The shield is manufactured from acrylic and bullet resistant materials to provide Level IIIA protection (9-mm and .44 magnum pistol rounds), weighs 3.4 pounds and is compatible with the PASGT helmet and M-17/M-40 protective masks.

Figure A-2. Riot Body Shield

3. Riot Body Shield: The riot body shield is a non-ballistic body shield designed to protect the service member's body from objects thrown or attacks from non-ballistic weapons. (Figure A-2)

No picture available.

4. Ballistic Body Shield with Light Kit: The ballistic shield provides the service member with torso protection from a wide range of threats (i.e. debris, liquids, hand thrown objects). The shield is manufactured from Spectra Shield resistant material painted anti-reflective flat black, and is capable of providing Level IIIA protection (9-mm and .44 magnum pistol rounds). The shield weighs 18 pounds and is 20 x 36 inches in size with a view port of 4 x 16 inches.

 **Figure A-3. Riot Shin Guard**	5. Riot Shin Guard: The riot shin guard is a non-ballistic shin guard designed to protect the service member's shins from objects thrown or attacks from non-ballistic weapons. (Figure A-3)
No picture available.	6. Ballistic Shin Guards: Ballistic shin guards provide the service member with leg protection from a wide range of threats (i.e. debris, liquids, hand thrown objects). The guards are manufactured from Kevlar KM2 and provide Level IIIA protection (9-mm and .44 magnum pistol rounds). They are available in three sizes and weigh between 7 to 10 pounds.
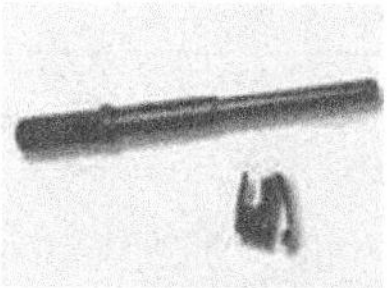 **Figure A-4. Expandable Riot Baton (with holster)**	7. Expandable Riot Baton (with holster): The expandable riot baton provides the service member a nonlethal means of crowd control and self-defense. The baton comes with a mounting device that attaches the baton to a belt. The baton is 24 to 36 inches in length. (Figure A-4)
No picture available.	8. Wooden Baton: The wooden baton provides self-defense for the service member and the option between physical force and lethal force. The baton comes with a mounting device that attaches the baton to a belt. The baton is 36 inches in length.
No picture available.	9. Practice Riot Batons: The practice riot batons provide the service member with a safe dynamic training environment for properly using a riot baton. The batons are available in a variety of lengths.
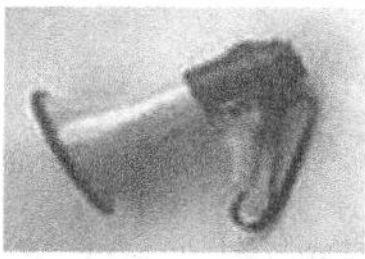 **Figure A-5. Portable Bullhorn**	10. Portable Bullhorn: The portable bullhorn provides the service member the capability to control his forces by projecting his voice above the noise and commotion created by crowds and mobs. (Figure A-5)
No picture available	11. Ground-Mounted Public Address System: The ground-mounted public address system is a communication enhancement device for conducting crowd control tactics. The system facilitates communication with the crowd in the absence of PSYOP support and assists with communication of commands to troops engaged in crowd-over-crowd noise.
No picture available.	12. M7 Individual Voice Amplification System: The M7 individual voice amplification system is a communication enhancement device for conducting crowd control tactics while wearing the M40 protective mask and using riot control agents. The M7 is fitted to the M40 protective mask to facilitate oral communication and increase the user's ability to communicate using radios and other devices.

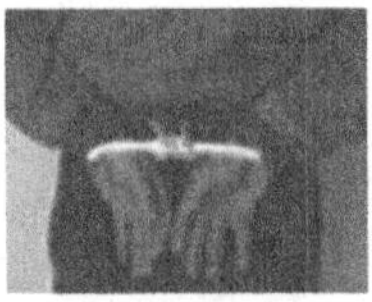

Figure A-6. Disposable Handcuff

13. Disposable Handcuff: The disposable handcuff provides service members with a means of restraining and controlling suspects, prisoners, and other individuals encountered during crowd control. The disposable handcuff is designed for one-time use. (Figure A-6)

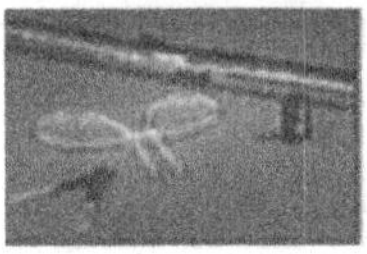

Figure A-7. Training Disposable Handcuff

14. Training Disposable Handcuff: The training disposable handcuff provides service members with a means for training in handcuffing procedures prior to actual use. The training disposable handcuff is red and reusable. (Figure A-7)

Figure A-8. MK-4 Pepper Spray Pouch

15. MK-4 Pepper Spray Pouch: The MK-4 pepper spray pouch is designed to carry the MK-4 pepper spray canister. (Figure A-8)

Figure A-9. MK-4 Individual Live OC Pepper Spray

16. MK-4 Individual Live OC Pepper Spray: The MK-4 individual pepper spray has a range of up to 12 feet in a target-specific stream. It contains 30 to 35 half-second bursts. (Figure A-9)

No picture available.

17. M-36 Individual Riot Control Disperser: M-36 Individual Riot Control Disperser has a range of up to 12 feet in a target-specific stream. It contains 30 to 35 half-second bursts. It is loaded with either CN or CS agent.

Figure A-10. MK-9 Pepper Spray Pouch

18. MK-9 Pepper Spray Pouch: The MK-9 pepper spray pouch is designed to carry the MK-9 pepper spray canister. (Figure A-10)

Figure A-11. MK-9 Squad Live OC Pepper Spray

19. MK-9 Squad Live OC Pepper Spray: The MK-9 squad pepper spray has a range of up to 15 feet in a target-specific fog. It contains 8 to10 half-second high volume bursts. (Figure A-11)

Figure A-12. MK-46 Platoon Live (Horizontal) OC Pepper Spray

20. MK-46 Platoon Live (Horizontal) OC Pepper Spray: The MK-46 platoon pepper spray has a range of up to 25 to 30 feet in a target-specific dispersed spray pattern. It contains up to 12 one-second high volume bursts. (Figure A-12)

Figure A-13. MK-46 Sling

21. MK-46 Sling: The MK-46 sling gives a service member the ability to sling the MK-46 canister on the shoulder. (Figure A-13)

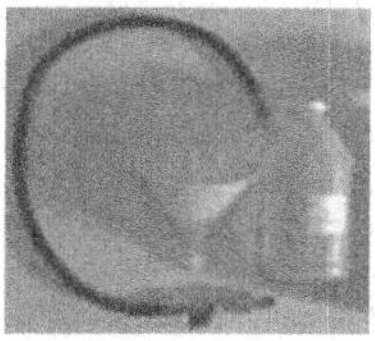

Figure A-14. MK-46 Hardware Kit

22. MK-46 Hardware Kit: The MK-46 hardware kit consists of a transfer tube, funnel, and transfer bottle. A compressed nitrogen source is also required to utilize the kit. (Figure A-14)

Figure A-15. MK-46 Refill Live OC Pepper Spray

23. MK-46 Refill Live OC Pepper Spray: The MK-46 refill solution comes in a one-gallon container. (Figure A-15)

Figure A-16. MK-4 Individual Inert OC Pepper Spray

24. MK-4 Individual Inert OC Pepper Spray: The MK-4 inert individual pepper spray has a range of up to 12 feet in a target-specific stream. It contains 30 to 35 half-second bursts. It provides the service member with an inert canister for familiarization and training purposes. The inert device is loaded with a non-irritant, non-toxic formulation payload. The device may also contain a non-toxic, water-soluble marker to allow immediate feedback on content delivery to target. The MK-4 device is marked "Inert." (Figure A-16)

Figure A-17. MK-9 Squad Inert OC Pepper Spray

25. MK-9 Squad Inert OC Pepper Spray: The MK-9 squad inert pepper spray has a range of up to 15 feet in a target-specific fog. It contains 8 to 10 half-second high volume bursts. It provides the service member with an inert canister for familiarization and training purposes. The inert device is loaded with a non-irritant, non-toxic formulation payload. The device may also contain a non-toxic, water-soluble marker to allow immediate feedback on content delivery to target. The inert MK-9 device is prominently marked "Inert" with a yellow sticker attached. (Figure A-17)

Figure A-18. MK-46 Platoon Inert OC Pepper Spray

26. MK-46 Platoon Inert OC Pepper Spray: The MK-46 platoon inert pepper spray has a range of up to 25 to 30 feet in a target-specific dispersed spray pattern. It contains up to 12 one-second high volume bursts. It provides the service member with an inert canister for familiarization and training purposes. The inert device is loaded win a non-irritant, non-toxic formulation payload. The device may also contain a non-toxic, water-soluble marker to allow immediate feedback on content delivery to target. The inert MK-46 device is prominently marked "Inert" with a yellow sticker attached. (Figure A-18)

Figure A-19. MK-46 Refill Inert OC Pepper Spray

27. MK-46 Refill Inert OC Pepper Spray: The MK-46 refill solution comes in a one-gallon container. (Figure A-19)

Figure A-20. FIST Training Suit

28. FIST Training Suit: The FIST training suit is made of closed-cell shock-absorbent foam that allows service members to hone their riot baton skills through realistic hand-to-hand engagement training. The FIST suit can absorb blows inflicted by the current expandable and wood riot baton. It provides protection for the head, face, hands, and legs, without significant degradation of the wearer's mobility. The FIST suit consists of a helmet, chest and back protector, bicep and forearm protectors, groin and buttock protectors, thigh and shin protectors, gloves, and carry bag. (Figure A-20)

NOTE: FIST is the current manufacturer of this training suit).

Figure A-21. Riot Training Strike Bag

29. Riot Training Strike Bag: The riot training strike bag gives the service member a strike bag for training and proficiency in open-hand control techniques. (Figure A-21)

Figure A-22. Caltrop

30. Caltrop: The caltrop is a field expedient area denial system. Securing approximately 50 caltrops to 550 military cord best employs it. This allows a service member to quickly toss/recover the system, using the caltrops to deny vehicle/pedestrian movement in a designated area. The four prongs are approximately two inches tall and will puncture tires or boots. (Figure A-22)

Figure A-23. Road Side Spike Strip

31. Road Side Spike Strip: The spike strip is designed to puncture a pneumatic tire as the vehicle makes contact with the spike strip. The spikes pivot into the tire and the tip-guards are pushed down. The spikes are imbedded deeply into the tire. The spikes are pulled out of the unit and remain in the tire. On the second revolution they will be pushed directly into the tire, allowing air to escape slowly without causing a blowout or sudden loss of control. The average tire will pick up 4 to 6 spikes with the deflation time of about 12 to 20 seconds. It allows a service member to quickly toss/recover the system, impeding vehicle movement in a designated area. It allows a force to provide additional physical security in areas where the caltrop cannot be tactically employed. (Figure A-23)

Figure A-24. High-Intensity Search Light

32. The high-intensity searchlight is a rugged, lightweight, hand-held, high-intensity spotlight for use during low-light conditions. Two different high-intensity lights were fielded. The Maxabeam light (shown on top) was fielded in 1997. It was replaced by the Nighthunter light (shown underneath) in 2000. (Figure A-24)

Figure A-25. Individual Light Holster (6P)

33. Individual Light Holster (6P): The 6P individual light holster is designed to hold the 6P individual light. (Figure A-25)

Figure A-26. 6P Individual Light

34. 6P Individual Light: The 6P individual light is issued to service members within the platoon. (Figure A-26)

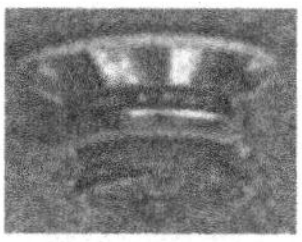

Figure A-27. 6P Individual Light Accessory Kit

35. 6P Individual Light Accessory Kit: The 6P individual light accessory kit includes a spare light bulb installation kit. (Figure A-27)

Figure A-28. 3V Lithium Batteries

36. 3V Lithium Batteries: Two 3V Lithium Batteries are required to operate the 6P individual light. (Figure A-28)

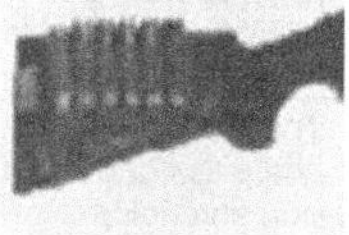

Figure A-29. 12-Gauge Buttstock Cuff

37. 12-Gauge Buttstock Cuff: The buttstock cuff is designed to fasten to the buttstock of most shotguns to hold seven 12-gauge cartridges for quick loading of the shotgun. (Figure A-29)

Figure A-30. 12-Gauge Utility Pouch

38. 12-Gauge Utility Pouch: The 12-gauge utility pouch is designed to hold twenty-five 12-gauge munitions and attaches to the service member's combat gear. (Figure A-30)

Figure A-31. 40-mm Carrying Pouch

39. 40-mm Carrying Pouch: The 40-mm carrying pouch is partitioned in a manner to provide the service member with a means of carrying not less than six 40-mm nonlethal rounds in separate compartments. It attaches to the service member's combat gear. (Figure A-31)

Figure A-32. Diversionary/Rubber Ball Grenade Pouch

40. Diversionary/Rubber Ball Grenade Pouch: The diversionary/rubber ball grenade pouch is an ammunition carrying pouch partitioned in a manner to provide the service member with a means of holding up to six rubber ball grenades or six flash bang grenades in separate compartments. (Figure A-32)

Figure A-33. 12-Gauge Mossberg 500A2/590 Shotgun

41. 12-Gauge Mossberg 500A2/590 Shotgun: The 12-gauge Mossberg 500A2 or 590 shotgun are the only 12-gauge shotguns currently authorized for firing 12-gauge nonlethal projectile munitions. The launching cup for the rubber ball grenade will attach to the Mossberg 500A2 Shotgun. The launch cup will not attach to the Mossberg 590 shotgun. Until the Joint Service Combat shotgun can effectively fire nonlethal 12-gauge rounds, the 12-gauge Mossberg 500A2 or 590 shotgun configuration meets the requirement of the NLW force. (Figure A-33)

Figure A-34. 12-Gauge M870 (modified) Version 2 (USCG)

42. 12-Gauge M870 (modified) Version 2 (USCG): The modified Remington 870 shotgun has a fixed synthetic stock and pistol grip. The pistol grip gives the shooter more control of the weapon and helps dissipate recoil from the shoulder. The synthetic materials used are more durable and lighter weight than the wood stocks used on the Standard M870. The barrel is fitted with a rifled choke tube for firing slugs (solid projectiles) to increase accuracy. It is also equipped with an Advanced Combat Optical Gunsight that was developed for use in "both-eyes-open" close quarters battle. The advantage is rapid target acquisition without the need to align front and rear sights. Batteries do not power the illumination sight; it contains tritium gas.

In the Coast Guard, a yellow "stock sock" will be used whenever the shotgun is loaded with nonlethal munitions. This removable nylon sleeve is used to identify the weapon as a nonlethal delivery system. (Figure A-34)

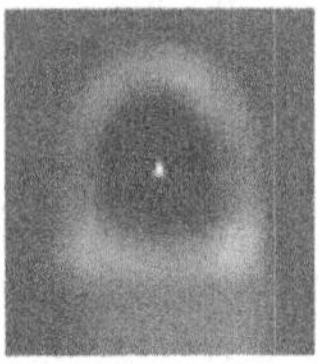

Figure A-35. Advanced Combat Optical Gunsight (ACOG) Reflex Sight (USCG)

43. Advanced Combat Optical Gunsight (ACOG) Reflex Sight (USCG): Modified shotguns are equipped with this site. (Figure A-34)

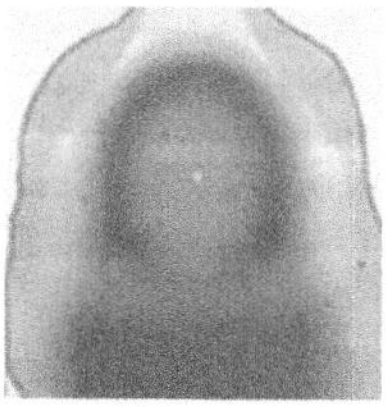

Figure A-36. ACOG Reflex Sight (USCG)

44. ACOG Reflex Sight (USCG): The dual-illuminated, Trijicon® Reflex sight was developed for the military for use in both-eyes-open close quarters battle (CQB). The advantage is quick target acquisition thus eliminating the need to align front and rear sights. Batteries do not power the illumination sight; it contains tritium gas. (Figure A-35)

Equipment Currently Being Fielded

1. 40-mm Crowd Dispersal Cartridge (CDC):

This direct fire, low-hazard, non-shrapnel-producing, multiple ball projectile provides friendly forces the capability to conduct crowd control. The CDC uses nonlethal, blunt trauma to stop, confuse, disorient, and/or temporarily incapacitate personnel and increase standoff distance between friendly forces and the hostile crowd. Primarily designed as an area effect munition, the CDC can be used on individuals. The M203 40-mm grenade launcher fires the CDC and contains 48, .48 caliber rubber balls. Effective range is 10 to 30 meters. (Figures A-37 and A-38)

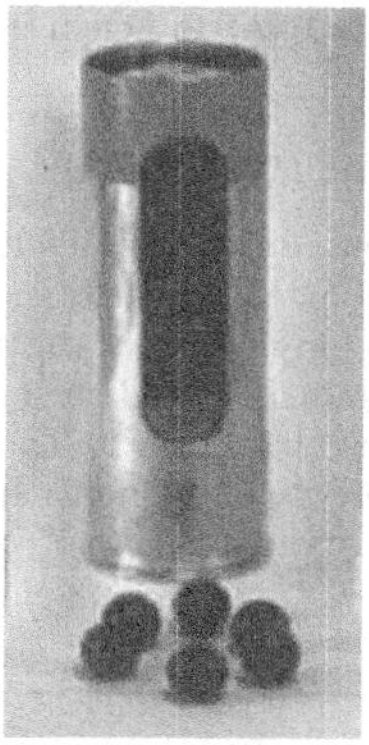

Figure A-37. Crowd Dispersal Cartridge

Figure A-38. Soldier Employing CDC with M203

2. Portable Vehicle Arresting Barrier (PVAB):

PVAB is a lightweight, portable, manually emplaced and recoverable barrier entrapment device to create, on command, a barrier that stops vehicles driven by known or suspected hostile forces with minimal damage to the vehicle or its occupants. It is used at dismount points, checkpoints, and roadblocks to prevent unauthorized access by wheeled vehicles into or out of areas (e.g. airfields, ports, command posts [CP]). Two individuals can easily emplace it, in approximately four hours. In stand-by mode, PVAB folds down into something similar to a speed bump (vehicles pass freely). In capture mode, the net springs up on two hydraulic pistons, to maximum height in less than a second. PVAB can be set up to span 12 to 24 feet and can stop a light truck (up to 7,500 pounds) at speeds up to 45 miles per hour, with a stopping distance of less than 200 feet PVAB can be containerized: 53.75 cubic feet (7 boxes, 4 large-3 small)—1,044 pounds on approximately one pallet. (Figures A-39 and A-40)

Figure A-39. PVAB in Action (Far View)

Figure A-40. PVAB in Action Near View

3. Modified Crowd Control Munition—Ground Emplaced (MCCM-GE):

The MCCM-GE (Figure A-41) is a nonlethal munition, housed in a casing similar to the current claymore mine, which will stop, confuse, disorient and/or temporarily incapacitate area targets/personnel at close range. The force can use it to provide crowd control by dispersing hostile crowds and force protection by integrating it into a layered perimeter defense. It delivers 600, .32 caliber rubber balls as well as a flash bang effect. The MCCM-GE gives the commander or small unit leader greater flexibility to apply nonlethal force as a first line of defense against hostile noncombatants. Effective range is 5 to 15 meters. (Figure A-42)

Figure A-41. Modified Crowd Control Munition-Ground Emplaced (MCCM-GE)

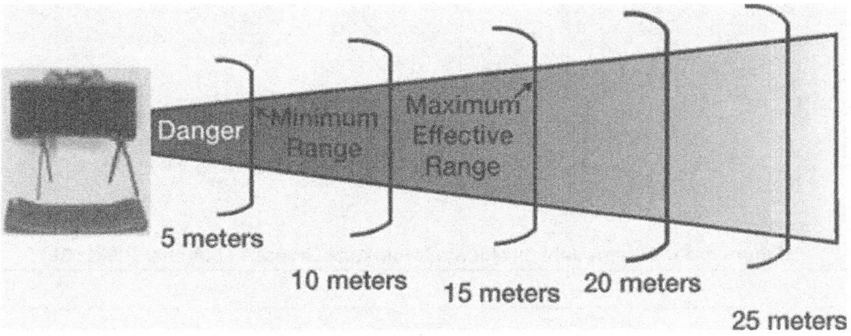

Figure A-42. MCCM-GE Range Chart

4. 66-mm Vehicle Mounted Nonlethal Grenade Launcher (VMNLGL):

The VMNLGL is a nonlethal grenade-launching device mounted on high mobility multipurpose wheeled vehicle (HMMWV). It is an indirect fire-support system that can deliver a 66-mm distraction device (flash-bang), CS, or blunt trauma (sting ball) payload. The distraction grenade uses a pyrotechnic charge for crowd control through audio and visual stimuli. The blunt trauma grenade uses a pyrotechnic charge to discharge 140, .32 caliber (cal.), rubber balls to achieve crowd control through audio, visual, and physical stimuli. The force uses VMNLGL to enhance the ability of friendly forces to conduct force protection without direct contact with hostile or threatening crowds. It enforces standoff distances and deters potential threats. It can disperse or subdue crowds, keep personnel from surrounding or climbing on patrol vehicles, slow or stop threatening personnel or vehicles by confusing, disorienting, and/or temporarily incapacitating individuals for capture. The grenade is fired from the grenade launchers traditionally found on tracked vehicles. Vehicle-launched grenades have three submunitions designed to ground burst. Effective range is 50 to 100 meters. An adjustable bracket allows engagement at ranges of 50 meters, 75 meters, and 100 meters. (Figure A-44)

Figure A-43. 66-mm Vehicle Mounted Nonlethal Grenade Launcher (VMNLGL)

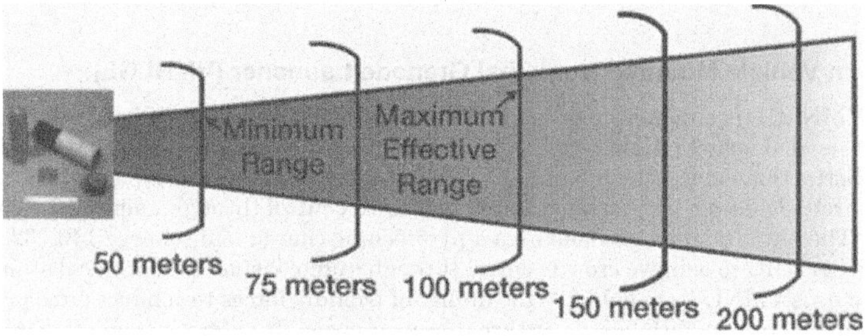

Figure A-44. VMNLGL Range Chart

5. Running Gear Entanglement System (RGES)

The Running Gear Entanglement System (RGES) is series of short, separate, looped polyethylene lines that are attached to a single, long polyethylene base line. This entanglement net fouls the propellers of planing hull small boats. The capability to manually deploy the net from a soft-sided bag has been fielded for USCG use in limited numbers. The USCG is in the process of incorporating three other variants of RGES into its inventory of nonlethal capabilities. These include aerial-delivered RGES, surface launched RGES to deploy from vessels, and static barrier RGES. The static barrier RGES has broad application into support homeland security, physical security and force protection missions. The surface launched RGES has direct application in the USCG's law enforcement mission wherever the requirement exists to stop a non-compliant vessel. Both of these systems are expected to be fielded in the fall/winter of CY02. The air launched version is currently still in the concept development phase and is not yet fully developed. (Figure A-45)

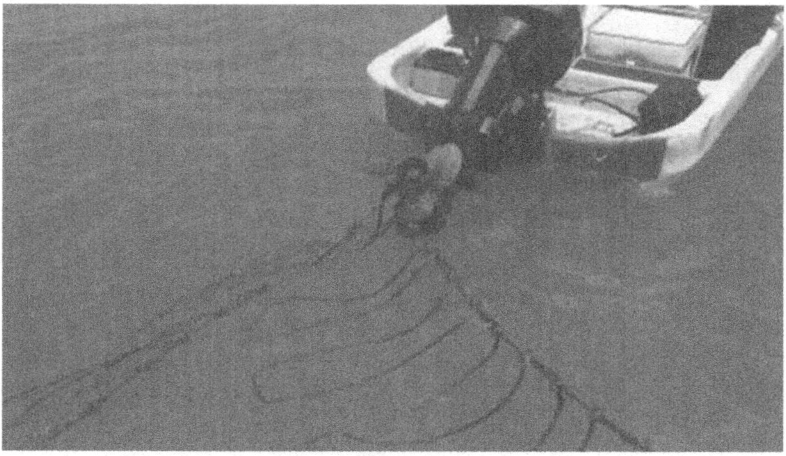

Figure A-45. Running Gear Entanglement System (RGES)

a. RGES Surface Delivery System: The USCG can use stern-mounted, manually deployed RGES from a cutter's small boat to stop slow-moving migrant-type craft. (Figure A-46)

(1) Uses COTS compressed air cylinder launcher.

(2) Removable, fix-mounted launcher on side of craft.

(3) Launch projects device ahead of target.

Figure A-46. RGES Surface Delivery System

b. RGES Barrier

(1) Provides static barrier around ship or other waterside asset.

(2) Provides a visible demarcation line, establishes zones, deters intrusions. (Figure A-47)

Figure A-47. RGES Barrier Employment

c. RGES Aerial Delivery System (ADS): RGES is aerial deployed from the ADS ahead of target aboard a H60 helicopter (Figure A-48). Device should float in the water ahead of target ensuring target runs over device. Device fouls propeller or rudder until target stops and awaits boarding.

(1) Uses COTS compressed air cylinder launcher

(2) Removable, fix-mounted launcher on helicopter

(3) Launch projects device ahead of target. (Figures A-49)

Figure A-48. RGES Aerial Delivery System (ADS)

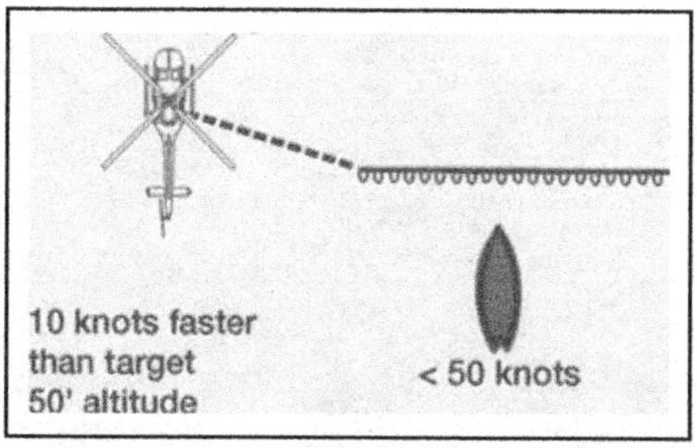

Figure A-49. RGES Aerial Delivery System (ADS)

6. Warp Cutter

a. Warp cutting is a nonlethal use of force, which has been safely and effectively employed by several countries to enforce fisheries laws in their respective exclusive economic zones (EEZ). The warp cutter is a device that closely resembles a grappling hook or small anchor. It has hardened steel blades attached to the crux of the grapnels. It is towed behind an enforcement vessel and across the warps of a non-compliant fishing vessel. As the cutting device engages the warps, it slices through them, separating the net from the fishing vessel.

b. The USCG uses the warp cutter to compel foreign vessels suspected of illegally fishing in the U.S. EEZ to comply with lawful orders to stop and be boarded. The USCG may cut the trawl warps of non-compliant vessels as a means of compelling them to stop for a boarding.

Figure A-50. Warp Cutter

Appendix B

Summary: Specialty Impact Munitions

12-Gauge Launch Cartridge

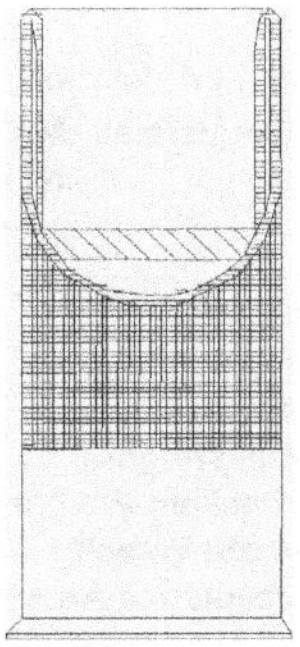

Using Services: USN, USMC

Status: Fielded 1998

DODIC: AA30

NSN: 1305-01-464-8389

Acquisition Activity: MARCORSYCOM—NSWC Crane, IN

Further Information:

Commercial: MARCORSYSCOM (703) 784-9395
NSWC Crane (812) 854-5801

Color .... Clear case
Diameter 12-gauge
Projectile .... N/A
Length .... 2.00 inches (51 mm)
Total Weight .... 13.6 grams
Propellant Weight .... 46.3 grains (black powder)
Muzzle Velocity .... 100 feet/second (30.5 m/sec)—rubber ball grenade
Acceptance Accuracy .... 200 feet (61 m) + 25 feet (7.6 m) at 30-degree launch angle using rubber ball grenade and 12-gauge launch cup

Projectile Specification: N/A

- Use 12-gauge launch round in combination with rubber ball grenade and 12-gauge launch cup to hurl a grenade a distance of 200 feet (61 m) + 25 feet (7.6 m) when launch angle is approximately 30 degrees.
- Chamber launch cartridge only after installing grenade in launch cup and pulling safety pin from grenade.

Figure B-1. 12-Gauge Launch Cartridge

12-Gauge Bean Bag Cartridge

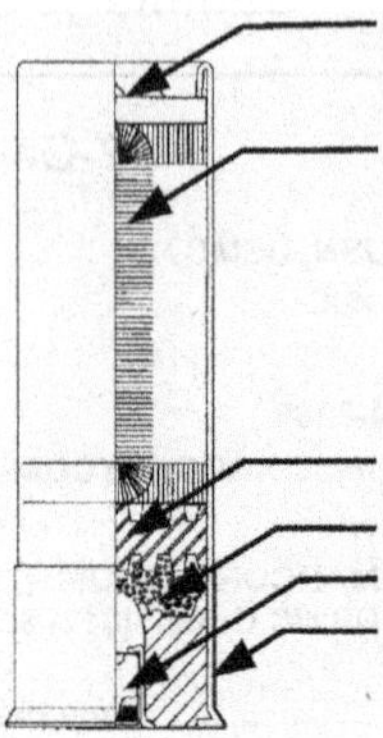

Using Services: USN, USMC
DODIC: AA29
NSN: 1305-01-454-0191
Status: Fielded 1998
Acquisition Activity: MARCORSYCOM—NSWC Crane, IN
Further Information:
Commercial: MARCORSYSCOM (703) 784-9395
NSWC Crane (812) 854-5801

Color....................................Clear or red case
Diameter................................12 gauge
Projectile ....1 bean bag
Length ....2.45 inches (62 mm)
Total Weight..........................49.9 grams
Propellant Weight.................3.4 grains
Muzzle Velocity.....................300 + 25 feet/second (91.4 + 7.62 m/sec)
Acceptance Accuracy..........2 feet (.6 m) X 6 feet (1.8 m) at 100 feet (30.3 m) with Mossberg 500A2

Shotgun Projectile Specification:
Shape...............................Square bag or tube sock
Material............................Nylon or cotton-polyester
Filler.................................#9 lead shot
Filler weight.....................44.4 grams (maximum)

- Fatal injuries are possible at employment distances of less than 20 feet (6 m)
- At distances of 20 feet (6 m) to 40 feet (12 m), target area should be center mass
- Training needs to reinforce that head shots are NOT ACCEPTABLE
- Do not skip fire this round! Round becomes unpredictable upon striking the ground
- Round is most effective against individually selected targets (point round)
- End wad has low light marker

Figure B-2. 12-Gauge Bean Bag Cartridge

12-Gauge Rubber Fin Stabilized Cartridge

Using Services: USN, USMC, USCG

Status: Fielded 1998

DODIC: AA31

NSN: 1305-01-454-0189

Acquisition Activity: MARCORSYCOM—NSWC Crane, IN

Further Information:

Commercial: MARCORSYSCOM (703) 784-9395
NSWC Crane (812) 854-5801

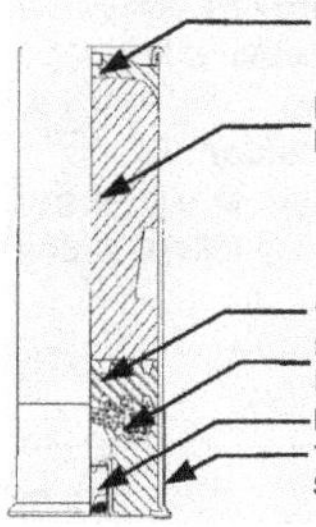

Color....................................Clear or red case
Diameter...............................12 gauge
Projectile...............................1 rubber projectile
Length...................................2.45 inches (62 mm)
Total Weigh...........................9.1 grams
Propellant Weight.................4.8 grains
Muzzle Velocity.....................530 + 65 feet/second
(161.5 + 19.8 m/sec)
Acceptance Accuracy..........2 feet (.6 m) X 6 feet (1.8 m) at 100 feet (30.3 m) with Mossberg 500 A2 shotgun
Projectile Specification: ......Caliber 12 gauge
Material.................................Rubber
Hardness...............................Durometer 70-80 Shore "A" scale
Weight...................................7.8 grams
Temperature Limits (demonstrated):
Upper limit...125° F
Lower limit...0° F

Shipping Data: Hazard classification/division and storage compatibility group...1.4S
DOT Class.............................ORM-D
DOT Markings Cartridge......Small Arms
UNO serial number...............0012

- **Fatal injuries are possible at employment distances of less than 20 feet (6 m); USCG: 15 yards**
- **At distances of 20 feet (6 m) to 40 feet (12 m), target area should be center mass; USCG: 15 to 30 yards**
- **Training needs to reinforce that head shots are NOT ACCEPTABLE**
- **Do not skip fire this round! Round becomes unpredictable upon striking the ground**
- **Round is most effective against individually selected targets (point round)**
- **End wad has low light marking**

Figure B-3. 12-Gauge Rubber Fin Stabilized Cartridge

40-mm Foam Rubber Baton Cartridge

Using Services: USN, USMC

Status: Fielded 1998

DODIC: BA07

NSN: 1310-01-453-9168

Acquisition Activity: MARCORSYSCOM—NSWC Crane, IN

Further Information:

Commercial: MARCORSYSCOM (703) 784-9395
NSWC Crane (812) 854-5801

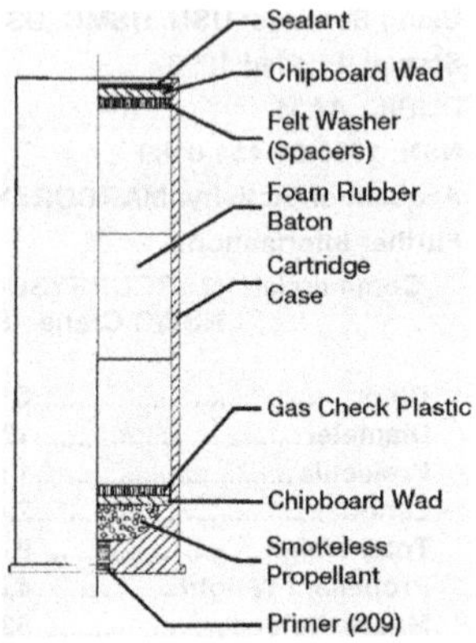

Color....................................Aluminum (case)
Diameter...............................40 millimeter
Projectile..............................3 foam batons
Length..................................4.78 inches (121.4 mm)
Total Weight..........................201 grams
Propellant Weight.................13.0 grains
Muzzle Velocity.....................325 + 35 feet/second (99 + 10.6 m/sec)
Acceptance Accuracy..........6 feet (1.8 m) X 3 feet (.9 m) at 50 feet (15.2 m) with M203

Projectile Specifications:

Number............................3
Material............................Foam rubber
Hardness.........................Durometer 25 + 10 on Shore "A" scale
Weight..............................250 to 280 grains

- Fatal injuries are possible at employment distances of less than 20 feet (6 m)
- At distances of 20 feet (6 m) to 40 feet (12 m), target area should be center mass
- Training needs to reinforce that head shots are NOT ACCEPTABLE
- Round is most effective against point targets (individual personnel)
- End wad has low light marker

Figure B-4. 40-mm Foam Rubber Baton Cartridge

40-mm Rubber Ball Cartridge

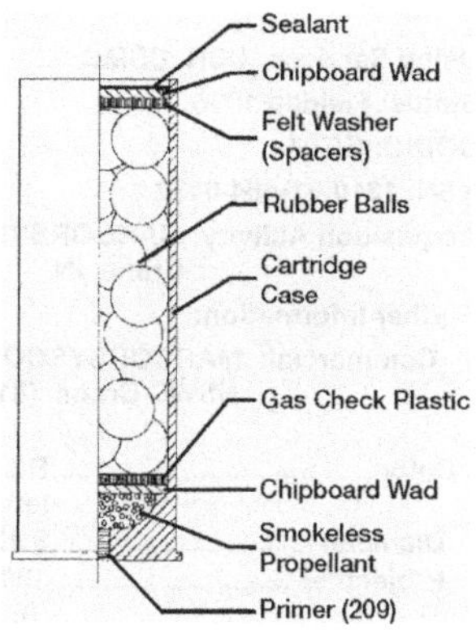

Using Services: USN, USMC

Status: Fielded 1998

DODIC: BA08

NSN: 1310-01-453-9154

Acquisition Activity: MARCORSYSCOM—NSWC Crane, IN

Further Information:

Commercial: MARCORSYSCOM (703) 784-9395

Color .... **Aluminum (case)**
Diameter .... **40 millimeter**
Projectile .... **24 rubber balls**
Length .... **4.78 inches (121.4 mm)**
Total Weight .... **201 grams**
Propellant Weight .... **7.3 grains**
Muzzle Velocity .... **325 + 35 feet/second (99 + 10.6 m/sec)**
Acceptance Accuracy .... **6 feet (1.8 m) x 9 feet (2.7 m) at 50 feet (15.2 m) with M203NSWC Crane (812) 854-5801**

Projectile Specifications:

Caliber **60**
Number .... **22 (minimum)**
Material .... **Rubber**
Hardness .... **Durometer 60 + 10 on Shore "A" scale**
Weight .... **2.2 grams each**

- **Fatal injuries are possible at employment distances of less than 20 feet (6 m)**
- **At distances of 20 feet (6 m) to 40 feet (12 m), target area should be center mass**
- **Training needs to reinforce that head shots are NOT ACCEPTABLE**
- **Round is most effective against area targets (multiple personnel)**
- **End wad has low light marker**

Figure B-5. 40-mm Rubber Ball Cartridge

Rubber Ball Grenade

Using Services: USN, USMC

Status: Fielded 1998

DODIC: GG04

NSN: 1310-01-454-0132

Acquisition Activity: MARCORSYSCOM—NSWC Crane, IN

Further Information:

Commercial: MARCORSYSCOM (703) 784-9395
NSWC Crane (812) 854-5801

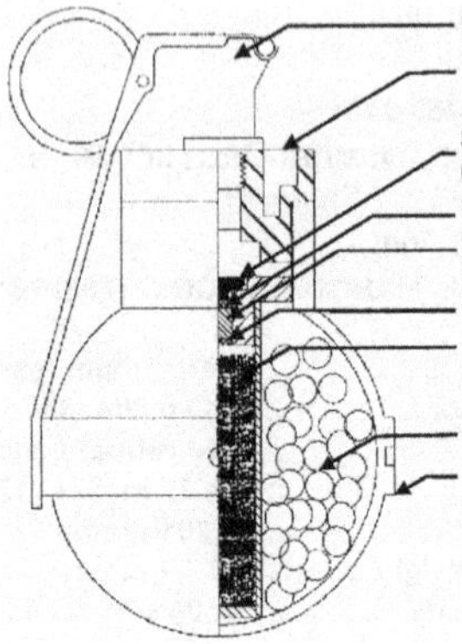

Color .... Black with white lettering
Diameter .... 3.12 inches (79 mm)
Projectiles .... 100 rubber balls (minimum)
Height .... 5.25 inches (133.4 mm)
Total Weight........................... 290 grams
Explosive Weight.................. 59 grams
Total Time Delay................... 3.0 + 0.5 seconds

Employment Distance:
Hand Thrown .... 50 feet (15.2 m)
Shotgun Launched .... 200 feet (61 m)

Projectile Specifications:
Number................................ 100 (minimum)
Material................................ Rubber
Hardness............................. Durometer 70 to 80 on Shore "A" scale
Diameter.............................. 0.25 inches (6.4 mm) (minimum)

- Fuze separates from the body 1.3 to 1.7 seconds after primer ignition.
- Second delay causes burster charge detonation 3.0 + 0.5 seconds average after primer ignition. Rubber balls are then dispersed in a 50-feet (15-m) circular pattern.
- 12-gauge shotgun with launch cup and launch cartridge can launch grenade 200 feet (61 m) + 25 feet (7.6 m)

Figure B-6. Rubber Ball Grenade

Practice Grenade Body

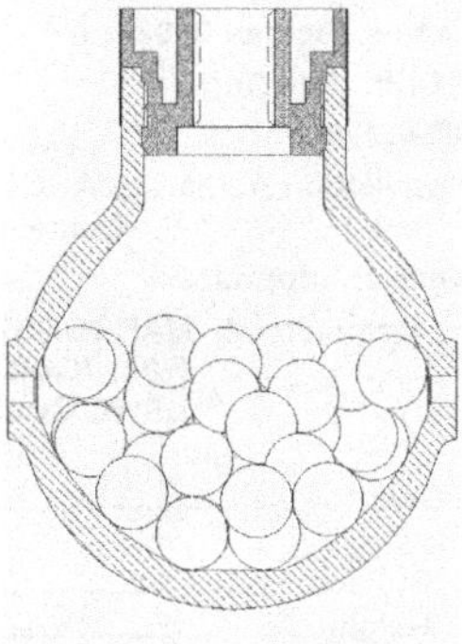

Using Services: USMC

Status: Fielded 2002

DODIC: GG05

NSN: Pending

Acquisition Activity: MARCORSYSCOM – NSWC Crane, IN

Further Information:

Commercial: MARCORSYSCOM (703) 784-9395
NSWC Crane (812) 854-5801

Diameter .... 3.12 inches (79 mm)
Projectiles .... N/A
Height .... 4.2 inches (106 mm)
Total Weight .... 208 grams
Explosive Weight .... N/A (inert)
Total Time Delay .... 3.0 + 0.5 seconds when used with practice grenade fuze

Employment Distance:
Hand Thrown .... 50 feet (15.2 m)
Shotgun Launched .... 200 feet (61 m) using Mossberg launch cup and 12-gauge launch cartridge (AA30).

- Practice grenade body is reusable up to 25 times and is a component of a practice grenade assembly, consisting of the body and a practice grenade fuze.
- Practice grenade assembly accurately replicates the GG04 rubber ball grenade and is used for training purposes.

Figure B-7. Practice Grenade Body

Practice Grenade Fuze

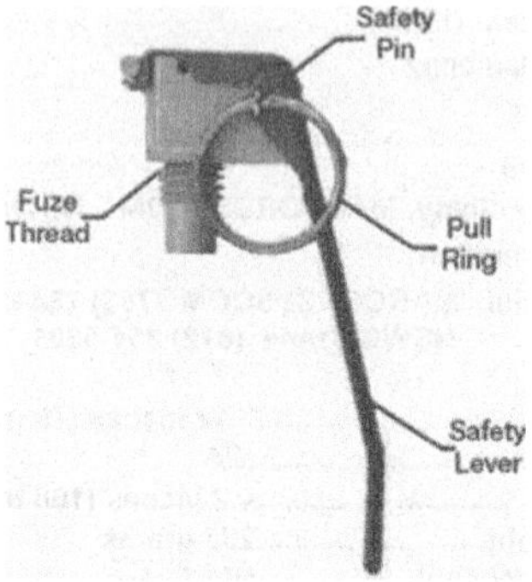

Using Services: USMC
Status: Fielded 2002
DODIC: Pending
NSN: Pending
Acquisition Activity: MARCORSYSCOM—NSWC Crane, IN
Further Information:
Commercial: MARCORSYSCOM (703) 784-9395
NSWC Crane (812) 854-5801

Color	Blue safety lever with brown stripe
Projectiles	N/A
Height	4.25 inches (107 mm)
Total Weight	82 grams
Explosive Weight	Not available
Total Time Delay	3.0 + 0.5 seconds

Employment Distance when used with GG05 Practice Grenade Body:

Hand Thrown	50 feet (15.2 m)
Shotgun Launched	200 feet (61 m)—using Mossberg launch cup and 12-gauge launch cartridge (AA30).

- Employ practice grenade fuze with the practice grenade body to create a practice grenade assembly consisting of the body and a practice grenade fuze.
- Practice grenade assembly accurately replicates the GG04 rubber ball grenade and is used for training purposes.

Figure B-8. Practice Grenade Fuze

Nonlethal Grenade Launch Cup

Using Service: USMC

Status: Fielded 2002

DODIC: N/A

NSN: Pending

Acquisition Activity: MARCORSYSCOM—NSWC Crane, IN

Further Information: Commercial: MARCORSYSCOM
(703) 784-2006 Ext. 2733
NSWC Crane (812) 854-5801

NOTE: Cup body shown is representative of cup bodies used on all shotguns. Adapter shown is subject to change depending on the shotgun used and design improvements.

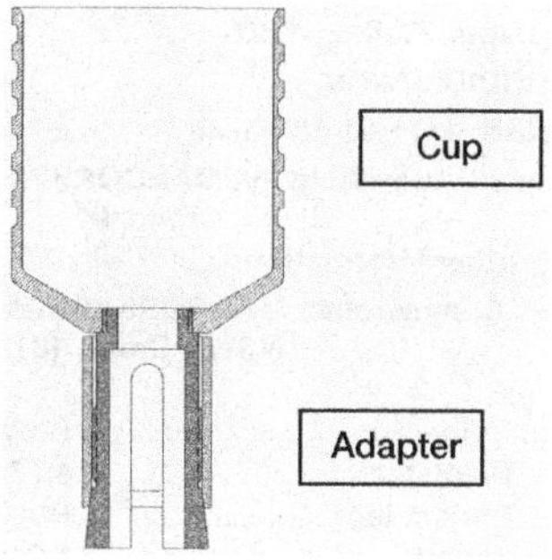

Cross Section View

Color	Black
Diameter (internal)	3.12 inches (79 mm)
Length (without adapter)	4 inches (102 mm)
Weight	397 grams (maximum)

- Use launch cup in concert with AA30 launch cartridge and either the GG04 rubber ball grenade or GG05 practice grenade assembly
- Attach launch cup and remove from the shotgun without using tools
- Launch cup can withstand accidental detonation of rubber ball grenade without fragmenting

Figure B-9. Nonlethal Grenade Launch Cup

Diversionary Charge, MK 141, MOD 0

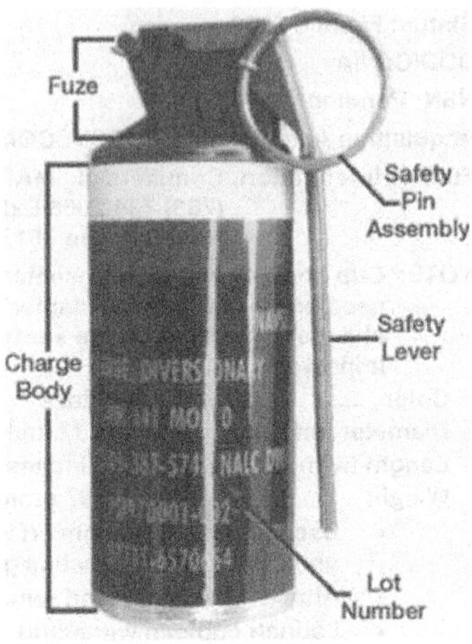

Using Services: USN, USMC
Status: Fielded 1997
DODIC: DWBS
NSN: 1375-01-387-5748
Acquisition Activity: MARCORSYSCOM—NSWC Crane, IN
Further Information:
Commercial: MARCORSYSCOM (703) 784-9395
NSWC Crane (812) 854-3934

Color	Brown with white stripe
Diameter	1.77 inches (45 mm)
Projectile	N/A
Length	5.214 inches (132.4 mm)
Total Weight	99.8 grams
Net Explosive Weight	17.8 grains
Light Output	2 million candlepower for 4 milliseconds
Noise Output	176 to 185 decibels
Material	Molded and machined polyurethane
Time Delay:	
Initial	1.26 seconds (minimum)
Total	1.34 to 1.84 seconds

- Diversionary charge can cause serious injury if held at time of detonation
- Light and noise most effective in confined space to stun individuals
- Most often employed for hostage rescue

Figure B-10. Diversionary Charge, MK 141, MOD 0

Cartridge, 12-Gauge Rubber Sting Ball™

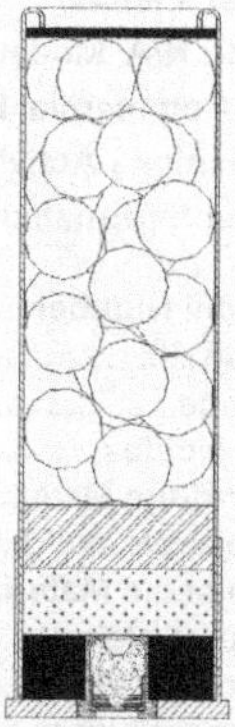

Using Services: USCG

Status: Fielded

DODIC: N/A Model No. 2552

NSN: Commercial Procurement

Acquisitions Activity: Commandant (G-OCU)

Further Information: Commandant (G-OCU) (202) 267-1522

Model Number	31-caliber sting-ball
Type	Multiple projectile rubber pellets
Range	10 yards (9 m)
Projectiles	18 (average)
Cartridge Material	Plastic
Projectile Weight	.35 grams each
Projectile Material	.31-caliber rubber pellets
Overall Height	2.75 inches (70 mm)
Overall Weight	18 grams
Compatibility	12-gauge shotguns
Velocity	400 feet/second

USCG

- Fatal injuries are possible at employment distances of less than 5 yards.
- At distances of 20 feet (6 m) to 40 feet (12 m), target area should be center mass
- This round is most effective against area targets with a maximum effective range of 15 yards
- Training needs to reinforce that head shots are NOT ACCEPTABLE
- Round is most effective against area targets (multiple personnel)
- Stock Sock: a yellow "stock sock" will be used whenever the modified 12-gauge Remington 870 shotgun is loaded with nonlethal munitions. This removable nylon sleeve is used to identify the weapon as a nonlethal delivery system.

- All specifications are average and are subject to change without notice
- 2552SPEC

Figure B-11. Cartridge, 12-Gauge Rubber Sting Ball™

Cartridge, 12-Gauge Sting Ball™ High Velocity

Using Services: USCG

Status: Fielded

DODIC: N/A Model No. 2553

NSN: Commercial Procurement

Acquisitions Activity: Commandant (G-OCU)

Further Information: Commandant (G-OCU) (202) 267-1522

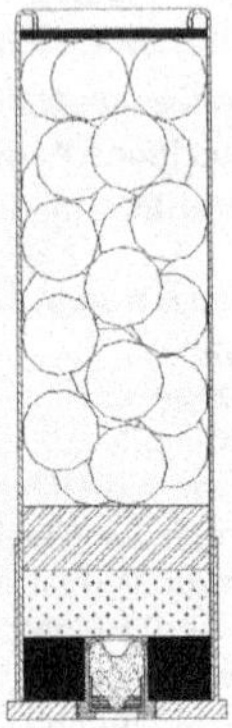

Model Number ....31-caliber sting-ball
Type ....Multiple projectile rubber pellets
Range ....10 yards (9 m)
Projectiles ....18 (average)
Cartridge Material.................Plastic
Projectile Weight ....35 grams each
Projectile Material..................31-caliber rubber pellets
Overall Height........................2.4 to 2.5 inches (61 – 63 mm)
Overall Weight ....18 grams
Compatibility..........................12-gauge shotguns
Velocity..................................900 feet/second

USCG

- Fatal injuries are possible at employment distances of less than 5 yards.
- At distances of 20 feet (6 m) to 40 feet (12 m), target area should be center mass
- Maximum effective range is 15 yards
- Training needs to reinforce that head shots are NOT ACCEPTABLE
- Round is most effective against area targets (multiple personnel)
- Stock Sock: a yellow "stock sock" will be used whenever the modified 12-gauge Remington 870 shotgun is loaded with nonlethal munitions. This removable nylon sleeve is used to identify the weapon as a nonlethal delivery system.

- All specifications are average and are subject to change without notice.
- 2553SPEC

Figure B-12. Cartridge, 12-Gauge Sting Ball™ High Velocity

M1006, 40-mm Nonlethal Cartridge (Sponge Grenade)

Using Services: U.S. Army

Status: Materiel Released April 2000

DODIC: BA06

NSN: 1310-01-452-1190

Acquisition Activity: Acquisition Center, Picatinny Arsenal, NJ

Further Information:

Commercial: Picatinny Arsenal NL Systems Integrator: (973) 724-6948
Product Director: (973) 724-6283

Color:

Projectile .... Black, olive green (foam rubber)

Cartridge Case .... Opaque

Diameter .... 40 millimeter

Length .... 5.272 inches (13.4 cm) (maximum)

Total Weight .... 68 grams (0.15 lbs)

Propellant Weight .... 2 grains

Muzzle Velocity .... 265 + 20 feet/second (81 + 6 m/sec)

Acceptance Accuracy .... >85 % accuracy at 30 meters (E-silhouette) and >60 % accuracy at 50 meters (E-silhouette) when fired from an M16A2/M203 or a M4/M203

Engagement Range .... 10 meters to 50 meters

Projectile Specification:

Shape .... 40-mm bullet shaped foam rubber

Material .... Foam rubber

Weight .... 28.5 grams (.06 lbs)

- Fatal injuries are possible at employment distances of less than 32.8 feet (10 m)
- At distances of 10 meters to 50 meters, target area should be center mass
- Training needs to reinforce that head shots are NOT ACCEPTABLE
- Do not skip fire this round! Round becomes unpredictable upon striking the ground
- Round is most effective against individually selected targets (point round)

Figure B-13. M1006, 40-mm Nonlethal Cartridge (Sponge Grenade)

M1012, 12-Gauge Shotgun: Nonlethal, Point Control Cartridge

Using Services: U.S. Army

Status: Type Classified, June 2000

DODAC: AA51

NSN: 1305-01-470-2405

Acquisition Activity: Acquisition Center, Picatinny Arsenal, NJ

Further Information:

Commercial: Picatinny Arsenal NL Systems Integrator: (973) 724-6948
Product Director: (973) 724-6086

Color:
Cartridge Case: .... Clear
Diameter .... 12 gauge
Length .... 2.45 inches (62.2 mm)
Total Weight .... 500 grains
Propellant Weight .... 6.6 grains smokeless (Red Dot)
Muzzle Velocity .... 500 feet/second (152.4 m/sec)
Acceptance Accuracy .... 80 % accuracy at 10 and 20 meters (E-silhouette) fired out of Mossberg 500, Mossberg 590, and Winchester 1200
Engagement range .... 10 to 20 meters
Projectile Specification:
Shape .... Fin stabilized 'bomblet' shape
Material .... Rubber
Weight .... 0.2 ounces (5.8 gm)
Hardness .... 75 Durometer "A" Scale

- Fatal injuries are possible at employment distances of less than 33 feet (10 m)
- At distances of 10 to 20 meters, target area should be center mass
- Training needs to reinforce that head shots are NOT ACCEPTABLE
- Round is most effective against individually selected targets (point round)

Figure B-14. M1012, 12-Gauge Shotgun: Nonlethal, Point Control Cartridge

M1013, 12-Gauge Shotgun: Nonlethal, Crowd Dispersal Cartridge

Using Services: U.S. Army

Status: Type Classified June 2000

DODAC: AA52

NSN: 1305-01-470-2139

Acquisition Activity: Acquisition Center, Picatinny Arsenal, NJ

Further Information:

Commercial: Picatinny Arsenal NL Systems Integrator: (973) 724-6948
Product Director :(973) 724-6086

Color:
- Projectile .... Black rubber
- Cartridge Case .... Clear

Diameter .... 12 gauge
Projectile .... 18 rubber balls
Length .... 2.37 inches (60.2 mm)
Total Weight .... 500 grains (32.4 gm)
Propellant Weight .... 14 grains smokeless (Red Dot)
Muzzle Velocity .... 900 feet/second (274 m/sec)
Acceptance Accuracy .... 80 % accuracy for a 2-by-3-meter target at 10 and 20 meters fired out of Mossberg 500, Mossberg 590, and Winchester 1200
Engagement range .... 10 to 20 meters
Projectile Specification:
- Shape .... 32-gauge (0.32 in) ball
- Material .... PVC rubber compound
- Weight .... 0.01 ounces (0.25 gm)
- Hardness .... 75 Durometer "A" Scale

- Fatal injuries are possible at employment distances of less than 32.8 feet (10 m)
- At distances of 10 to 20 meters, target area should be center mass
- Training needs to reinforce that head shots are NOT ACCEPTABLE

Figure B-15. M1013, 12-Gauge Shotgun: Nonlethal, Crowd Dispersal Cartridge

M1029, 40-mm Nonlethal Crowd Dispersal Grenade

Using Services: U.S. Army

Status: Type Classified May 2001

DODIC: BA13

NSN: 1310-01-475-0628

Acquisition Activity: Acquisition Center, Picatinny Arsenal, NJ

Further Information:

Commercial: Picatinny Arsenal NL Systems Integrator: (973) 724-6948
Product Director: (973) 724-6283

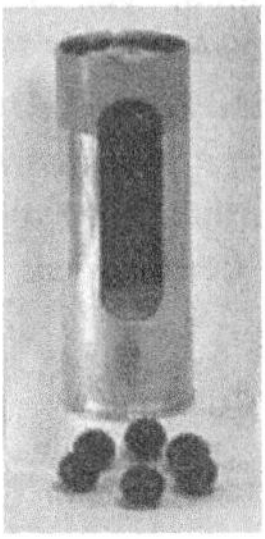

Color:

- Projectile .... Black rubber
- Cartridge Case .... Aluminum

Diameter .... 40 millimeter

Projectile .... 48 rubber balls

Length .... 4.8 inches (12.2 cm)

Total Weight .... 202.22 grams (7.0 oz)

Propellant Weight .... 8.5 grains smokeless

Muzzle Velocity .... 450 feet/second

Acceptance Accuracy .... > 90 % hit accuracy 4 out of 5 (shoulder to shoulder) E-silhouettes target at 30 meters and when fired from an M16A2/M203 or a M4/M203

Projectile Specification:

- Shape .... 48-gauge ball (.48 in diameter)
- Material .... Rubber
- Weight .... 1.3 grams (.046 oz)
- Hardness .... Durometer 60, scale "A"

- Fatal injuries are possible at employment distances of less than 32.8 feet (10 m)
- At distances of 10 to 30 meters, target area should be center mass of group of individuals
- Round is most effective when used to disperse groups of individuals

Figure B-16. M1029, 40-mm Nonlethal Crowd Dispersal Grenade

L96A1*, Grenade, Discharger: Antiriot, Irritant, CS

Using Services: U.S. Army, USMC
Status: Materiel Release, 4 January 2002
DODIC: FZ14
NSN: 1330-01-459-4018
Acquisition Activity: Acquisition Center, Picatinny Arsenal, NJ
Further Information:
Commercial: Picatinny Arsenal NL Systems Integrator: (973) 724-6948
Project Officer: (410) 436-2139

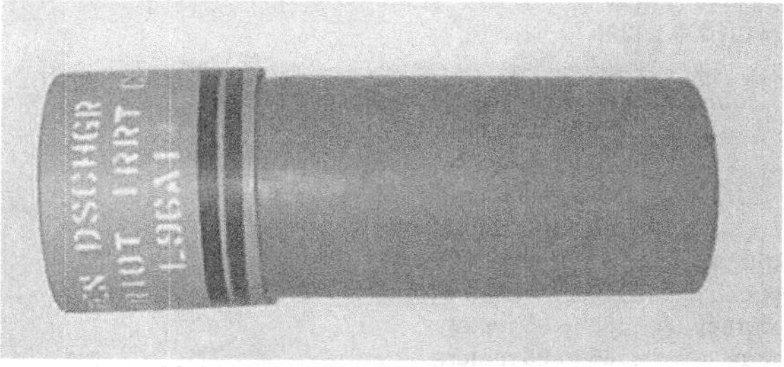

Color....................................Light gray with red and brown bands
Diameter................................66 millimeters
Projectile...............................23 canisters filled with CS compound
Length...................................185 millimeters (7.28 in)
Total Weight..........................568 grams (1.25 lbs)
Propellant Weight.................. 3.8 grams (.008 lbs)
Muzzle Velocity..................... 35.8 meters/second (117.5 ft/s)
Acceptance Accuracy..........65 to 95 meters
Deployment Range..............65 to 95 meters
Projectile Specification:

NOTE: It is not accurate to refer to the submunitions of the L96A1 grenade as projectiles. The antiriot material (CS) is released from the submunitions as they burn on the ground.

Shape...............................Cylindrical
Material............................Aluminum
Weight..............................11.3 grams (.025 lbs)

* The L97A1 Grenade is the training grenade for the L96A1 Grenade.

Figure B-17. L96A1, Grenade, Discharger: Antiriot, Irritant, CS

L97A1, Grenade, Discharger: Antiriot, Practice

Using Services: U.S. Army, USMC

Status: Material Released, 4 January 02

DODIC: FZ15

NSN: 1330-01-459-4032

Acquisition Activity: Acquisition Center,
Picatinny Arsenal, NJ

Further Information:

Commercial: Picatinny Arsenal NL Systems
Integrator: (973) 724-6948
Project Officer:
(410) 436-2139)

Color..Blue with brown and green bands
Diameter...................................66 millimeters
Projectile..................................23 canisters filled with Cinnamic Acid (CA)
Length.......................................185 millimeters (7.28 in)
Total Weight..............................568 grams (1.25 lbs)
Propellant Weight.....................3.8 grams (.008 lbs)
Muzzle Velocity.........................35.8 meters/second (117.5 ft/s)
Acceptance Accuracy..........65 to 95 meters
Deployment Range...............65 to 95 meters
Projectile Specification:

NOTE: It is not accurate to refer to the submunitions of the L97A1 grenade as projectiles. The antiriot, practice material (CA) is released from the submunitions as they burn on the ground.

Shape...............................Cylindrical
Material............................Aluminum
Weight..............................11.3 gram (.025 lbs)

- CA smoke can cause irritation to the eyes, mucous membranes, and skin. Personnel entering a CA cloud must wear appropriate face protection such as an M17- or M40-series protective mask.
- The burning grenade canisters could start a fire in dry grasses. Keep fire extinguisher available.
- During firing training, personnel within 16 meters must wear single hearing protection. This includes personnel within vehicle.
- Do not fire grenades when personnel or equipment are within 150 meters covering a 90-degree arc around a firing discharger.
- Do not fire grenades at the 50- or 75-meter position on the M315 adjustable aiming bracket. Fire only at the 100-meter setting.

Figure B-18. L97A1, Grenade, Discharger: Antiriot, Practice

Grenade, Launcher: Nonlethal, Distraction, M98

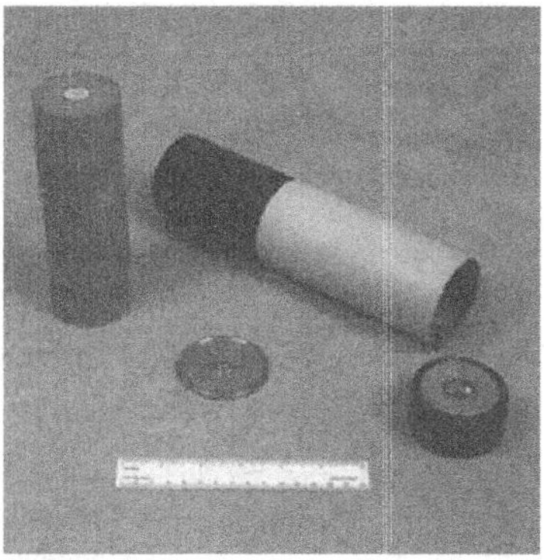

Using Services: U.S. Army, USMC

Status: Type Classified, 17 October 2001

DODIC: FZ16

NSN: 1330-01-484-7773

Acquisition Activity: Acquisition Center, Picatinny Arsenal, NJ

Further Information: Commercial: Picatinny Arsenal
NL Systems Integrator:
(973) 724-6948
Project Officer: (410) 436-2139

Color....................................Green and black with brown band
Diameter................................66 millimeters
Projectile...............................3 bursting submunitions, with pyrotechnic charges for audio and visual stimuli
Length...................................25.25 centimeters (9.94 in)
Total Weight..........................725.7 grams (1.6 lbs)
Propellant Weight.................5.0 grams (.011 lbs)
Muzzle Velocity.....................48.9 meters/seconds (160.4 ft/s)
Acceptance Accuracy..........100 to 120 meters
Deployment Radius..............10 to 15 meters
Projectile Specification:

NOTE: It is not accurate to refer to the submunitons of the M98 grenade as projectiles. These submunitions are ground bursting and the payload of inert ballast is just for flight stabilization and to deliver the audio/visual stimuli to the required range.

Filler................................Inert Ballast

- There is a remote chance of starting a fire. Keep fire extinguisher available.
- Personnel within 17 meters of a firing discharger must wear single hearing protection. This includes personnel within vehicle.
- Avoid looking directly at a bursting grenade; it could result in temporary loss of vision, lasting 2 to 5 minutes.
- Personnel within 35 meters of an operating launcher or 0.5 meters of a bursting canister should wear safety or ballistic-type eye protection, military clothing with long sleeves and elevated shirt collar, and helmet headgear.
- Do not handle or load M98s within 50 meters of a frequency transmitter emitting at 175 MHz, 448 MHz, 503 MHz and 1430 MHz.
- Do Not fire grenades when personnel or equipment are within 160 meters covering a 90-degree arc around a firing discharger.

Figure B-19. Grenade, Launcher: Nonlethal, Distraction, M98

M99, Grenade, Launcher: Nonlethal, Blunt Trauma

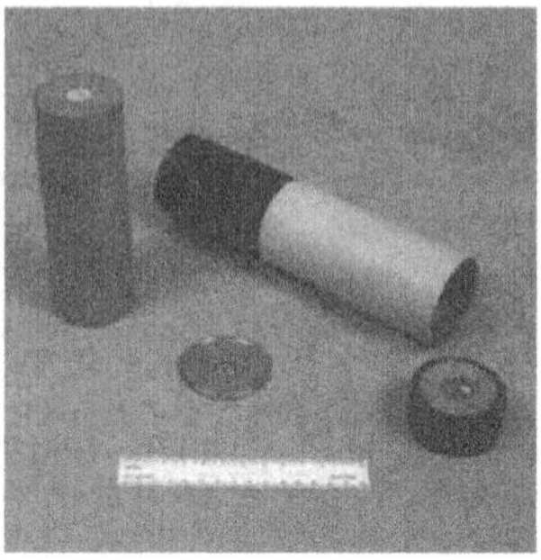

Using Services: U.S. Army, USMC

Status: Type Classified—17 October 2001

DODIC: FZ17

NSN: 1330-01-484-7775

Acquisition Activity: Acquisition Center, Picatinny Arsenal, NJ

Further Information:

Commercial: Picatinny Arsenal NL Systems Integrator: (973) 724-6948
Project Officer: (410-436-2139)

Color....................................Green and black with brown band
Diameter................................66 millimeters
Projectile...............................3 bursting submunitions, each with 140 balls
Length...................................25.25 centimeters (9.94 in)
Total Weight..........................725.7 grams (1.6 lbs)
Propellant Weight.................5.0 grams (.011 lbs)
Muzzle Velocity.....................48.9 meters/second (160.4 ft/s)
Acceptance Accuracy..........100 to 120 meters
Deployment Radius.............5 to 10 meters
Projectile Specification:

NOTE: It is not accurate to refer to the submunitions of the M99 grenade as projectiles. These submunitions are ground bursting and the payload of inert ballast is just for flight stabilization and to deliver the audio/visual stimuli to the required range.

Shape..............................32-caliber (5/16 in) ball
Material...........................PVC
Weight..............................0.40 grams (.014 oz)
Hardness.........................90 Durometer

- There is a remote chance of starting a fire. Keep fire extinguisher available.
- Personnel within 17 meters of a firing discharger must wear single hearing protection. This includes personnel within vehicle.
- Avoid looking directly at a bursting grenade; it could result in temporary loss of vision, lasting 2 to 5 minutes
- Personnel within 35 meters of an operating launcher or 0.5 meters of a bursting canister should wear safety or ballistic-type eye protection, military clothing with long sleeves and elevated shirt collar, and helmet headgear.
- Do not handle or load M98s within 50 meters of a frequency transmitter emitting at 175 MHz, 448 MHz, 503 MHz and 1430 MHz.
- DO NOT fire grenades when personnel or equipment are within 160 meters covering a 90-degree arc around a firing discharger.
- To minimize risk of eye injury, use the M99 primarily against crowds of adults.

Figure B-20. M99, Grenade, Launcher: Nonlethal, Blunt Trauma

Appendix C

Interservice Nonlethal Individual Weapons Instructor Course

1. Introduction

This appendix provides an overview of the Interservice Nonlethal Individual Weapons Instructor Course (INIWIC). The INIWIC is a "train-the-trainer" course taught in 15 subcourses and divided into four sections. Section I provides INIWIC descriptive data that includes the purpose, scope, peacetime and mobilization training requirements, student prerequisites, and equipment/ammunition requirements necessary to successfully complete the course. Section II lists the total training days and hours required per academic subcourse. Section III gives a detailed description of each subcourse and terminal learning objectives that each student must meet to complete the course. Section IV is a listing of 15 annexes that comprise the INIWIC Program of Instruction (POI). Access the entire INIWIC, with history, updates, and lesson plans at the following Internet address: http://www.wood.army.mil/mcdet/niwic/niwic.htm.

2. Course Descriptive Data

a. Course Title: Interservice Nonlethal Individual Weapons Instructor Course

b. Location: Marine Corps Detachment, Fort Leonard Wood, Missouri

c. Marine Corps service school code: A16H5A3

d. Other service course number: USAF—L5AZA3P071019

e. Military assistance program articles and service list number: P166810

f. Purpose: To certify selected DOD/DOT service persons as nonlethal individual weapons instructors.

g. Scope: This is a train-the-trainer course designed for E-4s and above. The objective of this course is to prepare students to train individuals and groups in the utilization and employment of NLW and to serve as the unit NLW subject matter expert. This course is designed for all occupational specialties.

h. Length: 10 training days

i. Curriculum Breakdown: 100 total hours.

(1) 2.00 Demonstration.

(2) 24.00 Demonstration/practical exercise.

(3) 22.50 Lecture.

(4) 5.00 Lecture/demonstration.

(5) 4.00 Live-fire exercise.

(6) 26.50 Practical application.

(7) 6.00 Performance exam.

(8) 4.00 Written exam.

(9) 6.00 Administrative time.

j. Maximum Class Capacity: 30.

k. Optimum Class Capacity: 25.

l. Minimum Class Capacity: 15.

m. Class Frequency: 12.

n. Prerequisites: Students must be selected by commanders to be nonlethal weapons instructors and be an E-4 or above.

o. MOS Received: None.

p. Quota Control: Formal Schools Branch, Training and Education Command, Quantico, Virginia.

q. Funding: Service or command funded.

r. Reporting Instructions: All students report with orders to the Fort Leonard Wood, Missouri, Billeting Office, Bldg # 470 no later than 2400 on the day prior to class convening. Government quarters are available; however, messing is not.

s. Instructor Staffing: Two USMC instructors and two U.S. Army instructors.

t. School Overhead: Course OIC and chief instructor

Training Support: Training support is provided by the USMC for all common training support items. Service-specific training items are provided by the respective service.

Table C-1. Required Equipment For INIWIC

Item	*Quantity*
Personnel Protectors:	
Non-Ballistic Riot Face Shield	40
Non-Ballistic Full Length Riot Shield	40
Non-Ballistic Shin Guards	40
Ballistic Riot Face Shield	20
Ballistic Full Length Riot Shield	20
Ballistic Shin Guards	20
Kevlar Helmets	40
Flak Jacket/Body Armor	40
Personnel Effectors:	
Disposable Handcuffs (Flex Cuffs)	300
12-Gauge Area (rubber ball)	1400
12-Gauge Beanbag	1400
12-Gauge Fin Stabilized	1400
12-Gauge Flash-bang	1400
12-Gauge OO Buck	xxxx
12-Gauge 7 1/2 shot	xxxx
12-Gauge Launching Cup Cartridge	1400
40-mm Area (rubber ball)	1400
40-mm Foam Baton (3 foam sub-munitions)	1400
40-mm Sponge Grenade	1400
MK141 Diversionary Grenade	600

Table C-1. Required Equipment For INIWIC

Item	*Quantity*
M84 Reloadable Diversionary Grenade	600
Sting Ball Hand Thrown or Shot launched Grenade	600
Mossberg 500A2 Shotgun Launch Cup	15
M203 Grenade Launcher	15
Mossberg 500 3" Shotgun	15
36" Riot Oak Baton w/Belt Ring	40
Monadnock® 24-36" Expandable Baton w/holster	40
M36 RCA Dispenser	40
Mk-4 Personal OC Dispenser	40
Mk-9 Team OC Dispenser	40
Mk-46 High Capacity High Output OC Dispenser	20
MK-46 Hardware Kit (Refill Kit)	10
M33A1 Squad Riot Control Agent Dispenser	10
Modular Crowd Control Munitions Ground Emplaced (MCCM-GE)	20
66-mm Vehicle Launched Nonlethal Grenade (VLNLG)	3
66-mm NL munitions	30
Mission Enhancers:	
Portable Bullhorn	5
40-mm Carrying Pouch	40
Diversionary Rubber Ball Grenade Pouch	40
12-Gauge Butt Cuff carriers	40
12-Gauge Utility 25-Round Pouch	40
Mk-4 Individual Carrying Pouch	40
Portable Public Address System w/Microphone	3
12-Volt Hand-Held Spotlight	5
Night Hunter High Intensity Light System w/Kit	5
12-Gauge High Intensity Light Set	5
Caltrops	5
Road Side Spike Strip	5
Portable Vehicle Arrest Barrier (PVAB)	3
Training Devices:	
Impact Training Suit	6
Striking Bags	15
Training Batons (24 in)	40
Training Batons (36 in)	40
Practice Grenade Body	40
Practice Grenade Fuse	540
• OPFOR Training Munitions (foreign objects)	xxxx
• Target Requirements (full or partial torso)	40
MK-4 Personal OC Dispenser (training)	300
MK-9 Team OC Dispenser (training)	300
Training Disposable Handcuffs (flex cuffs)	100
12-Gauge Dummy Rounds	100

Table C-2. INIWIC Class V Requirements

Type	*#Per Student*	*# Demo Rounds*	*Total Per Class*	*Total Per Year*
12-Gauge Area (rubber ball)	5	5	155	1400
12-Gauge Beanbag	5	5	155	1400
12-Gauge Fin Stabilized	5	5	155	1400
40-mm Area (rubber ball)	5	5	155	1400
40-mm Foam Baton	5	5	155	1400
Sting Ball Hand Thrown or Shotgun Launched Grenades	2	2	62	600
MK141 Diversion Devices	2	2	62	600

3. Training Days Summary

This course consists of 10 training days. The curriculum is broken down into three categories.

Table C-3. INIWIC Curriculum

NLW Tactics	*Hours*
Force Continuum/ROE	2.00
Crowd Dynamics/Control/Formations	14.00
Communication Skills	2.00
MOOTW	2.00
Tactical Considerations	2.00
	22.00
NLW Systems	***Hours***
Oleoresin Capsicum	4.00
Empty Hand Control Techniques	19.00
Riot Baton Techniques	8.00
Nonlethal Munitions and Equipment	8.00
Portable Vehicle Barrier	5.00
66-mm Grenade Launcher	2.00
Modular Crowd Control Munition	2.00
	48.00
Nonlethal Weapons Instructor	***Hours***
Techniques of Military Instruction	10.00
NLW Scenario Exercise	10.00
	20.00
Subtotal =	**90.00**
Examinations =	***4.00***
Administration =	***6.00***
Total Hours =	***100.00***

4. Course Body

The course body consists of 15 subcourses. A quick synopsis of each subcourse is included below.

a. **Instructor Development**. This subcourse teaches and reinforces basic military instructor methodology. It identifies characteristics of good instructors and four ways they can improve their teaching ability. Additionally, it illustrates rules of conducting military instruction as well as the steps in planning a lesson and teaching classes.

b. **Force Continuum**. This subcourse introduces the five levels of resistance as well as identifying the levels of the force continuum. It illustrates influencing the use of force decisions and the definitions relating to deadly force. Lastly, it identifies the three preconditions for deadly force and seven authorized occasions for use of deadly force.

c. **Crowd Dynamics/Crowd Control**. This subcourse outlines the differences between crowds, mobs, and riots, and teaches the student basic crowd control techniques to apply in various situations. Upon completion, the student will be able to instruct others in both the behavioral aspects of a crowd during a civil disturbance and also various techniques of controlling a civil disturbance. It will identify factors that affect both individual and crowd behavior and it identifies all crowd control equipment needed, or essential, for mission accomplishment.

d. **Empty Hand Control Techniques**. This subcourse identifies safety factors to consider when instructing empty hand control techniques by illustrating the four Ms (mind, mass, movement, muscle) of self-defense, psychomotor skills, and phases of psychomotor skills. It illustrates levels of training and identifies common pressure points of the human body. It identifies hinges, appropriate angles of approach, and recommended patterns of movement. Instructors will learn blocking techniques, striking skills, control holds, escort positions, and restraint devices.

e. **Baton Techniques**. This subcourse identifies riot baton nomenclature and safety standards when conducting riot baton training. It identifies methods of teaching psychomotor skills and the phases of teaching psychomotor skills. It highlights baton-striking techniques, blocking techniques, and retention techniques.

f. **Military Operations Other Than War (MOOTW)**. This subcourse defines operations other than war and the difference between MOOTW and declarations of war. It helps identify the primacy of political objectives and the strategic aspects of MOOTW. It illustrates the range of operations and the principles of operations, as well as the types of military operations other than war.

g. **Rules of Engagement (ROE)**. This subcourse defines ROE and specific terms that apply to these rules. It also defines the purpose of ROE and provides the student an understanding of policies underlying the need for ROE. It identifies three sources for ROE and the two parts of ROE that will not change. It explains how ROE relate to using and applying NLW.

h. **Oleoresin Capsicum (OC) Training**. This subcourse explains how the three levels of OC are created, as well as the nomenclatures for all aerosol canisters. It highlights the different spray patterns, methods of delivery, and cautions concerning the traumatic hydraulic needle effect. It explains employment considerations and the three levels of OC contamination. Additionally, it explains the physical and mental effects of OC and trains students on different grips, draw methods, and proper stances when employing OC. It explains procedures for decontamination, recovery, and first aid.

i. **Types of Formations**. This subcourse explains the elements and individual responsibilities within a riot control formation. Additionally, it trains proper grip, balance, holding, and retention of the riot control shield. It highlights leadership requirements by teaching different commands and cadence for conducting riot control formations. It illustrates the different types of formations and the support elements of a riot control operation

j. **Tactical Considerations**. This subcourse addresses the preparations necessary for crowd control operations and the graduated response of a crowd control force. It identifies support and information required from host nation police and civil authorities. It describes the employment measures of military forces in crowd control and MOOTW environments. It highlights the actions, optional measures, and guidance for using NLW in crowd control and MOOTW environments.

k. **Nonlethal Munitions and Employment**. This subcourse identifies the types of nonlethal munitions and their intended purpose. It illustrates how nonlethal munitions fall within established "use of force" continuums. It identifies the types, characteristics, and ranges of all currently employed nonlethal munitions.

l. **Communications Skills**. This subcourse identifies the areas of nonverbal communication employed to determine an individual's state of mind. This, along with an understanding of the importance of personal space, gives the student the tools to resolve conflict with the amount of force as applied in the force continuum.

m. **Portable Vehicle Arrest Barrier**. This subcourse teaches the employment and maintenance of a force protection asset assigned to the nonlethal capability set. The PVAB is designed to stop a 7,500-pound vehicle going 45 mph. The PVAB protects gates and check points.

n. **Modular Crowd Control Munition**. This subcourse teaches the employment of the MCCM; the nonlethal version of the claymore mine. The MCCM is a force protection asset designed to help protect static defensive positions or check points.

o. **M-315 66-mm Grenade Launcher**. This subcourse teaches the installation, employment, and maintenance of the M-315. The M-315 increases the current maximum range of nonlethal sting ball munitions from approximately 50 meters to 100 meters. Additionally, this weapons platform can deploy flash bangs, smoke, and CS munitions.

Appendix D

Sample: Unit Training Schedules/Firing Tables

Nonlethal Individual Weapons Training Course Training Schedule: Unit Training Schedule Number 1.

Training Day #1					
Time	***Subject***	***Location***	***Uniform***	***Instr***	***Ref/Notes***
0800	Administration Time	Classroom	U1		
0830	Force Continuum	Classroom	U1		
1000	ROE	Classroom	U1		
1200	Chow	Mess Hall	U1		
1300	MOOTW	Classroom	U1		
1500	Communication Skills	Classroom	U1		
1700	Secure/COB				

Administration:

Uniforms:

U1: BDU/utility uniform
U2: BDU/utility uniform with LBE/harness and canteens
U3: BDU/utility uniform with LBE/harness, canteens, flak jacket, and helmet
U4: BDU/utility uniform with LBE/harness, canteens, flak jacket, helmet, and assigned weapon

Transportation:

Weapons:

Ammunition:

Required Equipment:

Screen
Video projector with central processing unit
Televisions
VCR

Miscellaneous:

All personnel are required to have a notebook, pencil, and pen.
Handouts prepared and passed out.

Training Day #2					
Time	***Subject***	***Location***	***Uniform***	***Instr***	***Ref/Notes***
0800	Nonlethal Equipment Brief	Classroom	U1		
0900	Pressure Point Control Techniques/ Open Hand Control Techniques	Classroom	U1		
1000	Pressure Point Control Techniques/ Open Hand Control Techniques (Practical Application)	PT Field	U2		
1200	Chow				
1300	Pressure Point Control Techniques/ Open Hand Control Techniques (Practical Application)	PT Field	U2		
1700	Debrief/Secure/COB				

Administration:

Uniforms:
U1: BDU/utility uniform
U2: BDU/utility uniform with LBE/harness and canteens
U3: BDU/utility uniform with LBE/harness, canteens, flak jacket, and helmet
U4: BDU/utility uniform with LBE/harness, canteens, flak jacket, helmet, and assigned weapon

Transportation:

Weapons:

Ammunition:

Required Equipment:
Screen
Video projector with central processing unit
Flexicuffs (one set of cuffs per two students)
Riot Training Bags (one bag per two students)

Miscellaneous:
All personnel are required to have a notebook, pencil, and pen.
Handouts prepared and passed out.
Medical personnel on station.

Training Day #3					
Time	*Subject*	*Location*	*Uniform*	*Instr*	*Ref/Notes*
0800	Pressure Point Control	Classroom	U1		
	Techniques/Open Hand Control Techniques	Classroom	U1		
0900	Pressure Point Control Techniques/Open Hand Control Techniques (Practical Application)	PT Field	U1		
1200	Chow				
1300	Pressure Point Control Techniques/Open Hand Control Techniques (Practical Application)	PT Field	U2		
1700	Debrief/Secure/COB				

Administration:

Uniforms:

U1: BDU/utility uniform

U2: BDU/utility uniform with LBE/harness and canteens

U3: BDU/utility uniform with LBE/harness, canteens, flak jacket, and helmet

U4: BDU/utility uniform with LBE/harness, canteens, flak jacket, helmet, and assigned weapon

Transportation:

Weapons:

Ammunition:

Required Equipment:

Screen

Video projector with central processing unit

Flex cuffs (one set of cuffs per 2 students)

Riot Training Bags (one bag per 2 students)

Miscellaneous:

All personnel are required to have a notebook, pencil, and pen.

Handouts prepared and passed out.

Medical personnel on station.

Training Day #4					
Time	*Subject*	*Location*	*Uniform*	*Instr*	*Ref/Notes*
0800	Impact Weapons	Classroom	U1		
0900	Impact Weapons (Practical Application)	PT Field	U2		
1200	Chow				
1300	Impact Weapons (Practical Application)	PT Field	U2		
1700	Debrief/Secure/COB				

Administration:

Uniforms:

U1: BDU/utility uniform
U2: BDU/utility uniform with LBE/harness and canteens
U3: BDU/utility uniform with LBE/harness, canteens, flak jacket, and helmet
U4: BDU/utility uniform with LBE/harness, canteens, flak jacket, helmet, and assigned weapon

Transportation:

Weapons:

Baton (one baton per 2 students)
Training baton (one baton per 2 students)

Ammunition:

Required Equipment:

Screen
Video projector with central processing unit
Flex cuffs (one set of cuffs per 2 students)
Riot training bags (one bag per 2 students)
Three (3) riot training suits

Miscellaneous:

All personnel are required to have a notebook, pencil, and pen.
Handouts prepared and passed out.
Medical personnel on station.

Training Day #5					
Time	***Subject***	***Location***	***Uniform***	***Instr***	***Ref/Notes***
0800	Impact Weapons (Practical Application)	PT Field	U2		
1000	Crowd Dynamics/Crowd Control	Classroom	U1		
1130	Chow				
1230	Crowd Dynamics/Crowd Control	Classroom	U1		
1330	Riot Control Formations	Classroom	U1		
1430	Oleoresin Capsicum (OC) Aerosol Training	Classroom	U1		
1630	Debrief/Secure/COB				

Administration:

Uniforms:
U1: BDU/utility uniform
U2: BDU/utility uniform with LBE/harness and canteens
U3: BDU/utility uniform with LBE/harness, canteens, flak jacket, and helmet
U4: BDU/utility uniform with LBE/harness, canteens, flak jacket, helmet, and assigned weapon

Transportation:

Weapons:
Training baton (one baton per 2 students)
Baton (one baton per 2 students)

Ammunition:
Inert OC for classroom instruction

Required Equipment:
Riot training bags (one bag per 2 students)
Screen
Video projector with central processing unit
Three (3) riot training suits

Miscellaneous:
All personnel are required to have a notebook, pencil, and pen.
Handouts prepared and passed out.

Training Day #6					
Time	***Subject***	***Location***	***Uniform***	***Instr***	***Ref/Notes***
0800	OC Aerosol Training (Practical Application)	PT Field	U2		
1200	Chow				
1300	OC Aerosol Training (Practical Application)	PT Field	U2		
TBD	Debrief/Secure/COB				

Administration:

Uniforms:
U1: BDU/utility uniform
U2: BDU/utility uniform with LBE/harness and canteens
U3: BDU/utility uniform with LBE/harness, canteens, flak jacket, and helmet
U4: BDU/utility uniform with LBE/harness, canteens, flak jacket, helmet, and assigned weapon

Transportation:

Weapons:

Ammunition:
MK 4.inert OC (one per student)
MK 4 individual can OC (2 cans)

Required Equipment:
Five (5) large trash cans with liners
Two (2) water hoses with running water

Miscellaneous:
Medical personnel on station.

Training Day #7					
Time	*Subject*	*Location*	*Uniform*	*Instr*	*Ref/Notes*
0730	Movement to Range		U4		
0800	Nonlethal Munitions and Employment	Range Classroom	U1		
1030	Chow	Range			
1100	Safety Brief/Range Brief	Range	U4		
1130	Nonlethal Munitions (Live Fire)	Range	U4		
TBD	Evening Chow				
TBD	Safety Brief/Range Brief	Range	U4		
1800	Night Exercise Sun Set +30)	Range	U4		
TBD	Debrief/Clean Up	Range	U1		
TBD	Movement to Rear				
TBD	Secure/ COB				

Administration:

Uniforms:

U1: BDU/utility uniform
U2: BDU/utility uniform with LBE/harness and canteens
U3: BDU/utility uniform with LBE/harness, canteens, flak jacket, and helmet
U4: BDU/utility uniform with LBE/harness, canteens, flak jacket, helmet, and assigned weapon

Transportation: Yes

Weapons:

Assigned Weapons

Ammunition:

12-gauge rubber ball
12-gauge fin stabilized
12-gauge bean bag
12-gauge launching cup cartridge
40-mm foam baton
40-mm rubber ball
40-mm sponge grenade
Stingball grenades

Required Equipment:

Eight (8) 12-Gauge stinger launcher cup
Fifteen (15) targets
Eye and hearing protection
Fifteen (15) full-length riot shields
All students require munitions pouches, shin guards, face shields
Two (2) MRE per student
Video projector with central processing unit
Three (3) high-intensity light
Thirty (30) individual tactical lights

Miscellaneous:

Range safety officer.
Medical personnel on station.
Secondary means of communication (radios).
Conduct operational risk assessment.

Training Day #8					
Time	***Subject***	***Location***	***Uniform***	***Instr***	***Ref/Notes***
0700	Movement to Training Area #1		U2		
0730	Portable Vehicle Arresting Barrier (PVAB)	TA#1	U1		
0900	PVAB (Practical Application)	TA#1	U2		
1200	Chow				
1300	PVAB (Practical Application)	TA#1	U2		
1500	Debrief/Movement to Rear				
1530	Preventive Maintenance of PVAB				
1600	Secure/COB				

Administration:

Uniforms:
U1: BDU/utility uniform
U2: BDU/utility uniform with LBE/harness and canteens
U3: BDU/utility uniform with LBE/harness, canteens, flak jacket, and helmet
U4: BDU/utility uniform with LBE/harness, canteens, flak jacket, helmet, and assigned weapon

Transportation: Yes

Weapons:

Ammunition:

Required Equipment:
One (1) portable vehicle-arresting barrier (PVAB)
Video projector with central processing unit

Miscellaneous:
(1) MRE per student.

Training Day #9					
Time	***Subject***	***Location***	***Uniform***	***Instr***	***Ref/Notes***
0730	Movement to Range		U4		
0800	Modular Crowd Control Munition-Ground Emplaced (MCCM-GE)	Range Classroom	U1		
0900	Safety Brief/Range Brief	Range	U4		
0930	MCCM-GE (Practical Application	Range	U4		
1030	Chow				
1130	Vehicle Mounted Nonlethal Grenade Launcher (VMNLGL)	Range Classroom	U4		
1300	VMNGL	Range	U4		
1730	Debrief/Movement to Rear				
1800	Secure/COB				

Administration:

Uniforms:
U1: BDU/utility uniform
U2: BDU/utility uniform with LBE/harness and canteens
U3: BDU/utility uniform with LBE/harness, canteens, flak jacket, and helmet
U4: BDU/utility uniform with LBE/harness, canteens, flak jacket, helmet, and assigned weapon

Transportation: Yes

Weapons:

Vehicle Mounted Nonlethal Grenade Launcher (VMNLGL)

Ammunition:

(2) MCCM-GE, (2) 66-mm Riot Control Training Grenade (L97E1)

Required Equipment:
Eye and ear protection
Video projector with central processing unit

Miscellaneous:
Range safety officer.
Medical personnel on station.
Secondary means of communication.
Two (2) MRE per student.

Training Day #10					
Time	***Subject***	***Location***	***Uniform***	***Instr***	***Ref/Notes***
0700	Movement Down Range				U4
0730	Tactical Considerations	MOUT Classroom		U1	
0930	Safety Brief/Range Brief	MOUT Facility		U4	
1000	Tactical Exercise* (Practical Application)	MOUT Facility		U4	
1200	Chow				
1300	Tactical Exercise (Practical Application)	MOUT Facility		U4	
1700	Debrief/Movement to Rear				
1730	Weapons Maintenance				
1800	Secure/COB				

Administration:

Uniforms:
U1: BDU/utility uniform
U2: BDU/utility uniform with LBE/harness and canteens
U3: BDU/utility uniform with LBE/harness, canteens, flak jacket, and helmet
U4: BDU/utility uniform with LBE/harness, canteens, flak jacket, helmet, and assigned weapon

Transportation: Yes

Weapons:
Assigned weapon
Inert OC individual (one per student)
Training batons (one per student)
Fifty (50) practice grenades

Ammunition:

Required Equipment:
Three (3) riot training suit
All students require munitions pouches, flex cuffs, shin guards, face shields
Video projector with central processing unit
Fifteen (15) full length riot shields

Miscellaneous:
Range safety officer.
Medical personnel present.
Secondary means of communication.
(2) MRE's per student.

UNITED STATES MARINE CORPS
Marine Corps Detachment
Fort Leonard Wood, MO 65473

1.	LESSON DESIGNATOR:	Interservice Nonlethal Individual Weapons Instructor Course (INWIC). 13A (INIWIC.13A)
2.	LESSON TITLE:	Scenario Practical Exercise
3.	SCHOOL/COURSE:	Nonlethal Individual Weapons Instructor Course
4.	STUDY ASSIGNMENT:	Review course curriculum
5.	ESTIMATED TIME:	8 hours
6.	INSTRUCTORS REQUIRED:	(1) primary, (1) alternate per ratio: 1:10
7.	INSTRUCTOR REFERENCES:	INIWIC POI
8.	DATE PREPARED:	September 2001
9.	PRESENTATION METHOD:	Practical Exercise
10.	TRAINING AIDS:	Warning Order and five paragraph operations order for exercise, rubber rifles/shotguns, batons, shields, Inert MK4 OC canister, bullhorn, radio
11.	STUDENT EQUIPMENT:	(1) 782 gear (1) Kevlar with face shield (1) Shin guards

5. Unit Training Schedule Number 2

Table D-1. NLW Training Course 5-Day Training Schedule

Day/date	*Time*	*Event*	*Location*	*Instructor*	*Notes*
Mon/07 Jan 02	0730-0745	Welcome Aboard	Bldg RR202	MSgt Dixon	
T-1	0745-0845	Force Continuum	SOTG	TBD	L
	0900-1015	MOAB	SOTG	TBD	L
	1030-1130	OC	SOTG	TBD	L
	1130-1300	Noon Meal	Mess Hall	Company GySgt	
	1300-1400	OC Practical Application	SOTG	TBD	PA
	1400-1700	OC Certification	SOTG	TBD	PA
Tues/08 Jan 02	0730-0830	ROE	SOTG	TBD	L
T-2	0845-0945	Crowd Control	SOTG	TBD	L
	1000-1130	MDTS	SOTG	TBD	L, PA
	1130-1300	Noon Meal	Mess Hall	Company GySgt	
	1300-1700	MDTS	SOTG	TBD	L, PA
Wed/09 Jan 02	0730-1130	Batons	SOTG	TBD	L, PA
T-3	1130-1300	Noon Meal	Mess Hall	Company GySgt	
	1300-1500	Batons	SOTG	TBD	L, PA
	1500-1700	Formations	SOTG	TBD	L
Thurs/10 Jan 02	0730-1130	Formations	SOTG	TBD	PA
T-4	1130-1300	Noon Meal	Mess Hall	Company GySgt	
	1300-1700	Nonlethal Munitions/shotguns/ M2O3/sustainment training	Square bay/M203	TBD	L, PA
Fri/11 Jan 02	0730-1130	Nonlethal Munitions/stinger ball grenades/shotgun/M203	Square Bay/M203	TBD	PA
T-5	1130-1300	Noon Meal	Mess Hall	Company GySgt	
	1300-1700	Nonlethal Munitions/stinger ball grenades/shotgun/M203	Nonlethal Range 1&2	TBD	PA

Notes:

L = Lecture
PA = Practical Application
MDTS = Manadnock Defensive Tactics System
MOAB = Management of Aggressive Behavior
SOTG = Special Operations Training Group (USMC training unit)

6. Sample Unit Training Schedule Number 3

Table D-2. NLW Training Course 10-Day Training Schedule

DAY/DATE	TIME	EVENT	LOCATION	INSTRUCTOR	NOTES
Mon/29 Oct 01	0730-0745	Welcome Aboard	Bldg RR-202	MSgt Dixon	
T-1	0745-0845	Force Continuum	PT Field	Sgt Whitworth	L
	0900-1015	MOAB	PT Field	GySgt German	L
	1030-1130	OC	PT Field	Sgt Hoskins	L
	1130-1300	Noon Meal	Mess Hall	Company Gunny	
	1300-1400	OC Practical Application	PT Field	Sgt Hoskins	PA
	1400-1700	OC Certification	PT Field	Sgt Hoskins	PA
Tues/30 Oct 01	0730-0830	ROE	PT Field	Cpl Fewox	L
T-2	0845-0945	Crowd Control	PT Field	SSgt Parrish	L
	1000-1130	MDTS	PT Field	Sgt Johnson	L, PA
	1130-1300	Noon Meal	Mess Hall	Company Gunny	
	1300-1700	MDTS	PT Field	Sgt Johnson	L, PA
Wed/31 Oct 01	0730-1130	Batons	PT Field	Sgt Johnson	L, PA
T-3	1130-1300	Noon Meal	Mess Hall	Company Gunny	
	1300-1500	Batons	PT Field	Sgt Hoskins	L, PA
	1500-1700	Formations	PT Field	GySgt German	L
Thur/1 Nov 01	0730-1130	Formations	PT Field	Sgt Whitworth	PA
T-4	1130-1300	Noon Meal	Mess Hall	Company Gunny	
	1300-1700	Nonlethal Munitions	Nonlethal Range 1&2	GySgt German	L, PA
Fri/2 Nov 01	0730-1130	Nonlethal Munitions	Nonlethal Range 1&2	SSgt Parrish	PA
T-5	1130-1300	Noon Meal	Mess Hall	Company Gunny	
	1300-1700	Nonlethal Munitions	Nonlethal Range 1&2	SSgt Parrish	PA
Mon/5 Nov 01	0730-1130	Nonlethal Grenades Formation Fire	Nonlethal Range 1&2	GySgt German	L, PA
T-6	1130-1300	Noon Meal	Mess Hall	Company Gunny	
	1130-1700	Formation Fire/Sustainment Training	Nonlethal Range 1&2	SSgt Blackler	PA
Tues/6 Nov 01	0730-0830	Introduction to MOOTW	PT Field	Maj Desy	L
T-7	0830-1130	Vehicle Checkpoints/Vehicle Searches	PT Field	Sgt Whitworth	L, PA
	1130-1300	Noon Meal	Mess Hall	Company Gunny	
	1300-1500	Vehicle	PT Field	SSgt Parrish	PA

Table D-2. NLW Training Course 10-Day Training Schedule

DAY/DATE	*TIME*	*EVENT*	*LOCATION*	*INSTRUCTOR*	*NOTES*
		Checkpoints/Vehicle Searches			
	1500-1700	Roadblocks/Personnel Searches	PT Field	SSgt Blackler	L, PA
Wed/7 Nov 01	0730-1130	Roadblock/Personnel Searches	PT Field	SSgt Blackler	PA
T-8	1130-1300	Noon Meal	Mess Hall	Company Gunny	
	1300	Operations Order	Nonlethal Classroom	MSgt Dixon	Company Staff
Thurs/8 Nov 01	0730-1000	Confirmation Brief	PT Field	Maj Desy	
T-9	1000-1200	Insert	Combat Town	SSgt Parrish	
	1200-0000	STX	Combat Town	SSgt Parrish	
Fri/9 Nov 01	0001-1200	STX	Combat Town	SSgt Parrish	
T-10	1200-1300	Cleanup/Debrief	Combat Town	Company Gunny	
	1300	Graduation	Combat Town	SOTG OIC	

Notes:

L = Lecture

PA = Practical Application

MOAB = Management of Aggressive Behavior

MDTS = Manadnock Defensive Tactics System

STX = Situational Training Exercise

Appendix E

NLW Capability Sets

This appendix provides the current NLW capability sets for the services. These sets continue to be upgraded and changed according to service needs and the availability of new technologies. The U.S. Army and USMC capability sets are standard for the service, while the USAF, USN, and USCG are still developing and fielding their sets.

1. U.S. Army NLW Capability Set

Table E-1 U.S. Army Nonlethal Capability Set (Version FY 2001)

	Name	*Quantity*	*Unit Cost*	*Item Total*
GSAPMDK5X250AFHC	Non-Ballistic Riot Face Shields	200	$42.50	$8,500.00
GSAA TIPMBS9	Non-Ballistic Riot Body Shields Paulson	150	$88.50	$13,275.00
GSAATIUSMCSG	Non-Ballistic Riot Shin Guards	150	$60.00	$9,000.00
PT702MTUSA	Ballistic Face Shields	10	$246.00	$2,460.00
PT2237	Ballistic Body Shields	10	$1,052.00	$10,520.00
GSASF6SL	Sure Fire Field Light	10	$185.82	$1,858.20
MON2800 w/BCS-1	36" Hickory Riot Baton with Grommet	200	$15.50	$3,100.00
EADBH	Eagle Duty Baton Holder OD	200	$5.50	$1,100.00
GSAATIA2M	Portable Bullhorn	10	$82.00	$820.00
AAPB35W1	Ground Mount Bullhorn	8	$815.00	$6,520.00
AAPBWH10001	Megavox Wireless MIC	8	$0.00	$0.00
AASS450	Megavox Stand	8	$0.00	$0.00
GSASF6P	Individual High Intensity Light	15	$42.74	$641.10
GSAATIEATAVLFP6POD	Eagle Sure Fire 6P Holster	15	$8.75	$131.25
GSASFP60	Individual High Intensity Light Accessory Kit	15	$12.53	$187.95
XENNCSUSA	High Intensity Light	8	$2,795.00	$22,360.00
NIKFLEXCUFF	Disposable Forearm/Ankle Cuffs (Single)	1000	$0.70	$700.00
NIKCUTTER	Flex Cuff Cutters	2	$9.00	$18.00
GSAATIEAMK4	MK-4 Pouch OD	200	$6.75	$1,350.00
GSADEF5149	Inert Individual Training (OC) Dispenser	200	$6.07	$1,214.00
GSASF622	12-Gauge Shotgun High Intensity Light	50	$139.65	$6,982.50
GSAATIEAUPMS	12-Gauge Utility (25-Round) Pouch	50	$23.65	$1,182.50
GSAATISSSCQTR	12-Gauge Buttcuff Carrier (6-Round)	50	$10.75	$537.50
GSAATISASDD60D	Diversionary/Sting Ball Grenade Pouch	25	$47.25	$1,181.25
GSAATIEA40MM6	40-mm Carrying Pouch	100	$13.65	$$1,365.00
GSAATIRSC3	Caltrop	250	$10.50	$2,625.00
GSAATIFS333	Riot Training Suit with Accessories	4	$638.50	$2,554.00
GSAMONUTB	Riot Training Bag	10	$80.00	$800.00
BNCCELLS	C-Cell Batteries	0	$1.00	$0.00
BNAAACELLS	AA Batteries	0	$0.75	$0.00
GSASFDL123A	3V Lithium Batteries	0	$2.85	$0.00
BN9V	9V Batteries	0	$2.00	$0.00
ATIPMI	Packaging Marking and Integration Fee			$20,300.00
	Hardware Tools		**Item Total**	**$125,683.25**
	U.S. Army NLW Capability Set	**5**	**$125,683.25**	**$628,416.25**

2. USMC Nonlethal Weapons Capability Set

a. USMC NLW Company-Size Set

Table E-2 U.S. Marine Corps FY02 Company-Size Nonlethal Capability Set				
TAMCN C41202E	***NSN 9999-01-464-2285***	***SET A (Company Size)***		
NSN/Product ID	***Name***	***Quantity***	***Unit cost***	***Item Total***
GSAATINCSPMDK5	Non-Ballistic USMC Riot Face Shields (S/R)	200	$42.50	$8,500.00
GSAATIPMBS9	Non-Ballistic Riot Body Shields Paulson	60	$88.50	$5,310.00
GSAATIUSMCSG	Non-Ballistic Riot Shin Guards	200	$60.00	$12,000.00
GSAATI23/36	USMC 23"/36" Riot Baton (Expandable) (w/Holster)	200	$73.25	$14,650.00
GSAATITRAINING BATON	USMC Riot Baton (Practice) (Price TBD)	200	$20.00	$4,000.00
GSAATIA2M	Portable Bullhorn	12	$82.00	$984.00
GSAATIDC/W	USMC Disposable Handcuffs	400	$0.94	$376.00
GSAATIDC/T	USMC Disposable Handcuff Training Device	100	$0.94	$94.00
GSAATIEAMK4	MK-4 Pouch OD	200	$6.75	$1,350.00
GSADEF5049	MK-4 Individual Live (OC) Dispenser (Pepper Spray)	200	$7.29	$1,458.00
GSADEF5099	MK-9 Live Squad (OC) Dispenser (Pepper Spray)	40	$26.20	$1,048.00
GSATIEAMK9	MK-9 Pouch OD	40	$11.00	$440.00
GSAATIDEF5846L	MK-46 Live (OC) (Horizontal) (Pepper Spray)	12	$300.00	$3,600.00
GSAATIDEF1515	MK-46 Sling	12	$30.25	$363.00
GSAATIDEF5520	MK-46 Hardware Kit	4	$68.50	$274.00
GSAATIDEF5546	MK-46 Refill Unit (OC) Live (Pepper Spray)	4	$46.25	$185.00
GSADEF5149	MK-4 Inert Individual Training (OC) Dispenser	200	$6.07	$1,214.00
GSADEF5199	MK-9 Inert Squad Training (OC) Dispenser	40	$20.10	$804.00
GSAATIDEF5946L	MK-46 Inert Training (OC) (Horizontal) (w/Dye)	12	$240.00	$2,880.00
GSADEF5546I	MK-46 Refill Unit (OC) Inert Training (w/Dye)	4	$27.50	$110.00
GSAATIFS333	Riot Training Suit w/Accessories (F.I.S.T.)	12	$638.50	$7,662.00
GSAMONUTB	Riot Training Strike Bag	24	$80.00	$1,920.00
GSAATIRSC3	Caltrop	400	$10.50	$4,200.00
TBD	Stinger Road Side Spike Kit (15.5') (Price TBD)	12		
XENNLCSUSMC	High Intensity Light (Xenonics)	6	$2,975.00	$17,850.00
GSAATIEATAVLFP6POD	Eagle Sure Fire 6P Holster	60	$8.75	$525.00
GSASF6P	Individual High Intensity Light	60	$42.74	$2,564.40
GSASFP60	Individual High Intensity Light Accessory Kit	30	$12.53	$375.90
GSASFDL123A	3V Lithium Batteries	120	$2.85	$342.00
GSAATISSSCQTR	12-Gauge Buttstock Cuff	36	$10.75	$387.00
GSAATIEAUPMS	12-Gauge Utility (25-Round) Pouch	72	$23.65	$1,702.80
GSAATIEA40MM6	40-mm Carrying Pouch	36	$13.65	$491.40
GSAATISASDD60D	Diversionary/Sting Ball Grenade Pouch	60	$47.25	$2,835.00
CSI9512	12-Gauge Launching Cup (Availability and Price TBD)	36		
1005-01-371-5639/ 1005-01-251-8578	12-Gauge Mossberg 500A2 / 590 Shotgun	36	$250.00	$9,000.00
ATIPMI	Packaging Marking and Integration Fee			$11,578.00
	Hardware Totals		**Item Total**	**$121,073.50**

b. USMC NLW Platoon-Size Set

Table E-3 USMC FY02 NLW Capability Set—Platoon Size

TAMCN C41202E	*NSN 9999-01-464-2285*	*SET B (Platoon Size)*		
NSN/Product ID	***NAME***	***Quantity***	***Unit Cost***	***Item Total***
GSAATINCSPMDK5	**Non-Ballistic USMC Riot Face Shields (S/R)	50	$42.50	$2,125.00
GSAATIPMBS9	Non-Ballistic Riot Body Shields Paulson	15	$88.50	$1,327.50
GSAATIUSMCSG	Non-Ballistic Riot Shin Guards	50	$60.00	$3,000.00
GSAATI23/36	USMC 23"/36" Riot Baton (Expandable) (w/Holster)	50	$73.25	$3,662.50
GSAATITRAINING BATON	USMC Riot Baton (Practice) (Price TBD)	50	$20.00	$1,000.00
GSAATIA2M	Portable Bullhorn	3	$82.00	$246.00
GSAATIDC/W	USMC Disposable Handcuffs	100	$0.94	$94.00
GSAATIDC/T	USMC Disposable Handcuff Training Device	25	$0.94	$23.50
GSAATIEAMK4	MK-4 Pouch OD	50	$6.75	$337.50
GSADEF5049	MK-4 Individual Live (OC) Dispenser (Pepper Spray)	50	$7.29	$364.50
GSADEF5099	MK-9 Live Squad (OC) Dispenser (Pepper Spray)	10	$26.20	$262.00
GSATIEAMK9	MK-9 Pouch OD	10	$11.00	$110.00
GSAATIDEF5846L	MK-46 Live (OC) (Horizontal) (Pepper Spray)	3	$300.00	$900.00
GSAATIDEF1515	MK-46 Sling	3	$30.25	$90.75
GSAATIDEF5520	MK-46 Hardware Kit	2	$68.50	$137.00
GSAATIDEF5546	MK-46 Refill Unit (OC) Live (Pepper Spray)	2	$46.25	$92.50
GSADEF5149	MK-4 Inert Individual Training (OC) Dispenser	50	$6.07	$303.50
GSADEF5199	MK-9 Inert Squad Training (OC) Dispenser	10	$20.10	$201.00
GSAATIDEF5946L	MK-46 Inert Training (OC) (Horizontal) (w/Dye)	3	$240.00	$720.00
GSADEF5546I	MK-46 Refill Unit (OC) Inert Training (W/Dye)	2	$27.50	$55.00
GSAATIFS333	Riot Training Suit W/Accessories (F.I.S.T.)	3	$638.50	$1,915.50
GSAMONUTB	Riot Training Strike Bag	3	$80.00	$240.00
GSAATIRSC3	Caltrop	50	$10.50	$525.00
TBD	Stinger Road Side Spike Kit (15.5') (Price TBD)	3		
XENNLCSUSMC	High Intensity Light (Xenonics)	4	$2,975.00	$11,900.0 0
GSAATIEATAVLFP6POD	Eagle Sure Fire 6P Holster	50	$8.75	$437.50
GSASF6P	Individual High Intensity Light	50	$42.74	$2,137.00
GSASFP60	Individual High Intensity Light Accessory Kit	25	$12.53	$313.25
GSASFDL123A	3V Lithium Batteries	100	$2.85	$285.00
GSAATISSSCQTR	12-Gauge Buttstock Cuff	15	$10.75	$161.25
GSAATIEAUPMS	12-Gauge Utility (25 Round) Pouch	30	$23.65	$709.50
GSAATIEA40MM6	40-mm Carrying Pouch	50	$13.65	$682.50
GSAATISASDD60D	Diversionary/Sting Ball Grenade Pouch	50	$47.25	$2,362.50
CSI9512	12-Gauge Launching Cup (Availability and Price TBD)	15		
1005-01-371-5639/ 1005-01-251-8578	12-Gauge Mossberg 500A2 / 590 Shotgun	15	$250.00	$3,750.00
ATIPMI	Packaging Marking and Integration Fee			$3,000.00
	Hardware Totals		**Item Total**	**$43,471.25**

** Riot shield. Marine Corps security forces are currently evaluating to use in conjunction with each other as listed below. They are not currently deployed but are undergoing field evaluation.
(1) The six feet Armadillo Interlocking Shield for the hard baseline.
(2) The four feet.. Armadillo Interlocking Shield for the soft baseline.

3. USN NLW Capability Set

a. The USN is in the process of developing NLW capabilities and capability sets.* The sets will support the specific areas of operation, mission, and/or user.

b. The first version will address the force protection mission and will be based on other service-approved items and some longer-range commercial-off-the-shelf items currently being upgraded for military use. The mobile security forces, master at arms, and shipboard force protection units are the targeted users.

c. A notional capability set for the Mobile Security Forces is listed below.

Table E-4. Notional USN Capability Set

Name	*Nomenclature (Mfr. Representative)*	*Quantity*	*Unit Cost*	*Case Size*	*Cost Extended*
Fin Stabilize Round		1 Case	$5	100	$400
Stingshot		1 Case	$5	100	$400
Shotbag		1 Case	$5	100	$450
Flamethrower Round		1 Case	$8	100	$750
Tactical Blast Bursting Round		1 Case	$7	100	$650
Ultraflash Bursting Round		1 Case	$8	100	$750
Launchable Stun Munition	POP, Inc. T-444	1 Case	$60	24	$1,440
40-mm Sparkshot		1 Case	$35	24	$840
40-mm Stingshot		1 Case	$25	24	$600
40-mm Shotbag		1 Case	$25	24	$600
40-mm OC		1 Case	$28	24	$660
Stinger	POP, Inc. I-452	1 Case	$60	24	$1,440
Stun	POP, Inc. T-470	1 Case	$55	24	$1,320
Flash/Bang Multi	POP, Inc. D-451	1 Case	$65	24	$1,560
OC	POP, Inc. I-452	1 Case	$65	24	$1,560
12-Gauge Launching Blanks		200	$3		$600
Grenade Training Rounds		12	$65		$780
12-Gauge Training Rounds		24	$3		$72
40-mm Training Rounds		12	$20		$240
12-Gauge Shotguns	Standard Issue/Modified	12	$1,000		$12,000
12-Gauge Launching Adapter		6	$325		$1,950
Face Shields		15	$100		$1,500
Helmets		15	$200		$3,000

* In FY 01, the USN fielded an USMC capability set to Naval Air Station, Roosevelt Roads, Puerto Rico.

Table E-4. Notional USN Capability Set

Name	*Nomenclature (Mfr. Representative)*	*Quantity*	*Unit Cost*	*Case Size*	*Cost Extended*
Tactical Vest		15	$300		$4,500
Bullhorn		2	$100		$200
Medical Kit		2	$50		$100
Searchlights		2	$100		$200
Field Pack		15	$100		$1,500
Holsters/Belts		15	$50		$750
Ordnance Packaging, Handling, Storage and Transportation (PHS&T)					$1,500
Misc. (Certification, Training and Documentation)					$3,000
			Total Cost Basic Kit		**$45,312**

Table E-5. Notional USN Capability Set

Name	*Nomenclature (Mfr. Representative)*	*Quantity*	*Unit Cost*	*Case Size*	*Cost Extended*
Stingmore Mine (Std Body) with Rubber Ball Projectiles	POP, Inc. I-454	24	$350		$8,400
Stingmore Mine Stand		6	$100		$600
Electric Firing Boxes		8	$35		$280
Area Lighting					$1,000
Electric Generator					$500
Signs and Placards					$500
Personal NVGs and Opticals		6	$2,000		$12,000
Additional PHS&T					$2,000
Misc. (Certification, Training and Documentation)					$2,000
Total Cost Augmented Basic Kit					**$27,280**
Total Cost (Basic plus Augment)					$72,592
Portable Vehicle Arresting Barrier	U.S. Army	1	$40,000		$40,000
Running Gear Entanglement System (RGES)	NSWCDD	1000 ft system	$25,000		$25,000
Antimaterial Round (USCG)		24	$12		$288
Spike Strips		2	TBD		
Improved Sensor (s)					
RF			TBD		

Table E-5. Notional USN Capability Set

Name	*Nomenclature (Mfr. Representative)*	*Quantity*	*Unit Cost*	*Case Size*	*Cost Extended*
IR			TBD		
Acoustic			TBD		
Chemical Detection	SABRE 2000	1	$20,000		$20,000
Ordnance Packaging, Handling, Storage and Transportation (PHS&T)					$6,000
Misc. (Certification, Training and Documentation)					$2,000
Total Cost Vehicle Supplement					**$93,288**
Total Cost Basic, Augmented, and Vehicle Supplement					**$165,880**

4. USAF NLW Capability Set

Table E-6. USAF NLW Capability Set

NSN	*Item Description*	*EA*	*Item $*	*Item Total*
GSADEF5149	Inert Individual (OC) Training Dispenser	1	$6.07	$6.07
GSADEF5846H	Mid-Size Riot Control Dispenser (Chemical)	3	$318.75	$956.25
GSADEF5546	Riot Control Dispenser (OC) Agent Quarts	1	$46.25	$46.25
	12-Gauge Shotgun (Area CDC/Rubber Pellet)	36	$4.46	$160.56
	12-Gauge Shotgun (Point/Fin Stabilized)	36	$2.85	$102.60
CSI (USMC) GG04	Stun Hand (Sting Ball) Grenade	50	$35.00	$1,750.00
M84	Stun Hand Grenade/Flash-Bang Device	50	$78.00	$3,900.00
GSAPMDK5X/AF	Non-Ballistic Face Shields	13	$29.50	$383.50
GSAPMBS9	Non-Ballistic Full Length Riot Shield	13	$88.50	$1,150.50
GSAUSMCSG	Non-Ballistic Shin guards	13	$6.000	$780.00
GSAATI23/36	36" Riot Baton w/Holster	13	$73.25	$952.25
GSAA2	Portable Megaphone	1	$82.00	$82.00
EAAFCUSTOM	Tactical Vest Rough Estimate	3	$150.00	$450.00
HARG700	Riot Control Gloves	13	$52.20	$678.60
Champro	Protective Cups	13	$7.25	$94.25
	40-mm (Point) Sponge Round	54	$23.00	$1,242.00
	40-mm (Area) Crowd Dispersal Cartridge Grenade	54	$22.00	$1,188.00
GSAF333	Riot Baton Training Suit	1	$638.50	$638.50

Table E-6. USAF NLW Capability Set

NSN	***Item Description***	***EA***	***Item $***	***Item Total***
MONDC/W	Disposable Forearm/Ankle Cuffs (FLEX)	40	$0.94	$37.60
GSADEF5049/ EA	Individual (OC) Dispenser with Carrier	13	$7.29	$94.77
	12-Gauge Utility Pouch	2	$24.00	$48.00
	12-Gauge Buttstock Cuff	3	$11.00	$33.00
	40-mm Carrying Pouch	2	$14.00	$28.00
	Training Strike Bag	2	$80.00	$160.00
	Cost of Clip-Lok Boxes			$480.00
	Approximate freight shipped via commercial freight into CONUS			$1,500.00
			Total Estimate	**$16,942.70**
Proposal to add TASERS to USAF Capability Set after testing.				

5. Notional USCG NLW Capability Set

Table E-7. Notional USCG NLW Capability Set

Reference: (a) COMDTINST M16247.1B Maritime Law Enforcement Manual (MLEM) (b) Project Ericson Website	
Expandable Baton	
Restraint Devices	Hand Cuffs, Flex Cuffs, Leg Irons, Migrant Restraint Harness
Individual OC Spray	
Non-Ballistic Riot Face Shield	
Non-Ballistic Riot Body Shield	
Non-Ballistic Riot Shin Guard	
Running Gear Entanglement Systems	

The following information describes each of the components of the NLW capability set currently deployed by the USCG. (The USCG has not designated a standard set of nonlethal gear. The using unit determines the gear needs.)

6. Restraint Devices

a. Handcuffs are the preferred restraint devices when dealing with a small number of subjects. Handcuffs should be steel, chain-linked construction capable of being double locked, with nickel, blued, or stainless steel finish.

b. Using flexible handcuffs for the temporary restraint of subjects is also authorized. Since they do not require a key and are relatively easy to apply, flexible handcuffs are particularly useful and appropriate when dealing with large numbers of subjects (e.g. migrants).

c. Leg irons are authorized restraint devices and may be a preferred option for restraining subjects for extended periods of time.

d. Migrant restraint harnesses are authorized when dealing with large numbers of subjects.

7. Expandable Baton

The expandable baton is the only intermediate weapon authorized for USCG personnel to carry. It should have a steel shaft that is silver in color and corrosion resistant (e.g. chrome, nickel, stainless steel) and have no knurling or sharp edges. It should have a foam or similar, non-slip grip. It should be rigid when deployed and must be no less than 15 inches and no greater than 22 inches when expanded. Spring loaded or other trigger style batons are not authorized.

8. OC Pepper Spray

Only chemical irritants containing the active ingredient OC are authorized for USCG personnel to carry. The aerosol must contain no greater than 10 percent and no less than 5 percent OC concentration. It shall have a water-based carrier, be nonflammable, nontoxic and produce a straight stream delivery (not fog).

9. Helmet Riot Face Shield

Comparable to DOD type. Not currently deployed. Undergoing field evaluation.

10. Riot Shield

Two shields are currently being evaluated. A 24 x 48 inches clear lexan weighing 14 pounds and a smaller version measuring 20 x 36 inches. They are not currently deployed. Undergoing field evaluation.

REFERENCES

Joint

JP 1-02, *DOD Dictionary of Military and Associated Terms*, 12 Apr. 2001

JP 3-07, *Joint Doctrine for Military Operations Other Than War*, 16 Jun. 1995

JP 3-07.3, *JTTP for Peace Operations*, 12 Feb.1999

JP 3-57, *Doctrine for Joint Civil-Military Operations*, 8 Feb. 2001

JP 3-58, *Joint Doctrine for Military Deception*, 31 May 1996

JP 3-61, *Doctrine for Public Affairs in Joint Operations*, 14 May 1997

JP 4-0, *Joint Doctrine for Logistic Support of Joint Operations*, 6 Apr. 2000

Multi-Service

FM 3-100.12, MCRP 5-12.1C, NTTP 5-03.5, AFTTP(I) 3-2.34, *Risk Management* for *Multiservice Tactics, Techniques, and Procedures*, 15 Feb. 2001

FM 3-31.1, MCWP 3-36, *Army and Marine Corps Integration in Joint Operations*, 21 Nov. 2001

Army

AR 190-14, *Military Police Carrying of Firearms and Use of Force for Law Enforcement and Security Duties*, 12 Mar. 1993

AR 350-38, *Training Devices Policies and Management*, 15 Oct. 1993

AR 27-1, *Judge Advocate Legal Services*, 30 Sep. 1996

FM 3-0, *Operations*, 14 Jun. 2001

FM 3-05.30 (33-1), *Psychological Operations*, 19 Jun. 2000

FM 19-15, *Civil Disturbances*, 25 Nov. 1985

FM 27-10, *The Law of Land Warfare*, 18 Jul. 1956

FM 100-19, *Domestic Support Operations*, 1 Jul. 1993

TC 90-1, *Training for Urban Operations*, 1 Apr. 2002

TRADOC Pamphlet 525-5, *Force XXl Operations, A Concept for the Evolution of Full Dimensional Operations for the Strategic Army of the Twenty-First Century,* 1 Aug. 1994

TRADOC Pam 350-9, *TRADOC Training Devices for Army wide Use*, 30 Sep. 1993

TRADOC Regulation 350-70, *Systems Approach to Training Management, Processes, and Products*, 9 Mar. 1999.

TRADOC Regulation 385-2, chg.1, TRADOC Safety Program, 10 Oct. 2000.

Marine Corps

MCWP 3-33.4, *Domestic Support Operations*, Jul. 1993

I MEF FMF, After Action Report—UNITED SHIELD, 7 Jun. 1995

MCO 1510.112, *Individual Training Standards for Nonlethal Weapons*, 27 May 1998

MCO 5500.6F, *Arming of Security and Law Enforcement Personnel and the Use of Force*, 20 Jul. 1995

MCRP 5-12.1A, *The Law of Land Warfare*.

MCO 1630.4A, *Law Enforcement and Physical Security Activities*, 19 Jun. 1991.

MCO 1510.114, *Individual Training Standards (ITS) for Antiterrorism Force Protection (AT/FT) System*, 30 Jan. 1998.

MCWP 3-11.1, *Marine Rifle Company*, Feb. 1978

MCRP 3-11.1A, *Commander's Tactical Handbook*, 30 Jul. 1988. .

MCWP 3-11.2 *Marine Rifle Squad*

Marine Corps Gazette, *"Less-Lethal" Force in Operation UNITED SHIELD*, Sep. 1995.

Navy

Naval Doctrine Publication 1, Naval Warfare (U), 28 Mar. 1994

NWP 1-14M, *The Commander's Handbook on the Law of Naval Operations*, Oct. 1995

SECNAVINST 5500.29B, *Use of Deadly Force and the Carrying of Firearms by Personnel of the Department of the Navy in Conjunction with Law Enforcement Security Duties and Personal Protection*, 29 Sep. 1992

Air Force

AFI 31-201, *Security Police Standards and Procedures*, 4 Dec. 2001

AFI 31-207, *Arming and Use of Force by Air Force Personnel*, 1 Sep. 1999

AFMAN 31-201V6, *Civil Disturbance*, 17 May 2002

AFMAN 31-222, *Security Forces Use of Force Manual*, 1 Jun. 2001

Coast Guard

COMDTINST M16247.1B, Maritime Law Enforcement Manual, 9 Aug. 2001

Other

DODD 3000.3-R, *Policy for Nonlethal Weapons*, 9 Jul. 1996

DODD 3025.1, *Military Support to Civil Authorities (MSCA)*, 15 Jan. 1993

DODD 5100.77, *DOD Law of War Program*, 9 Dec. 1998

DODD 3025, *I-M Manual for Civil Emergencies*, 2 Jun. 1994

DODD 3025.12, *Military Assistance for Civil Disturbances*, 4 Feb. 1994

DODD 3025.13, *DOD Resources in Support of US Secret Service*, 13 Sep. 1985

DODI 5000.2, *Operation of the Defense Acquisition System*, 5 Apr. 2002

DODR 5000.1, *Defense Acquisition*, 23 Oct. 2000, Ch 1, 4 Jan. 2001

DOD Joint Concept for Nonlethal Weapons, 5 Jan. 1998

DODD 3025.12, *Military Assistance for Civil Authorities*, 18 Feb. 1997

DODD 3025.16, *Military Assistance for Civil Authorities Liaison Office (EPLO) Program*,18 Dec. 2000

DODI 5000.2, *Operation of the Defense Acquisition System*, 5 Apr. 2002

Mission Need Statement (MNS), *Range of Nonlethal Force Capabilities (NLFC*), 10 Oct. 1996

A Guide for Effective Peacetime Employment of Reserve Components Units and Individuals, OASD RA (M&P), Dec. 2000

Defense and Intelligence Abbreviations and Acronyms, Nov. 1997

Emergency Management Assistance Compact—Guide Book and Operating Procedures, National Emergency Management Association, Aug. 2000

Inside the Navy, Marine Corps Nonlethal Weapons Experiment Underway at Camp Pendleton, CA, Vol. 9, No. 29, Jul. 22, 1996

Military Support To Civil Authorities—Interagency Disaster Resource Guide, National Interagency Civil-Military Institute, Jan. 2001

National Guard Bureau Military Support to Civil Authorities Lessons Learned, 6 Sep. 2000

The Robert T. Stafford Disaster Relief and Emergency Assistance Act, as amended 42 USC 1521, Jul.2001

Joint Universal Lessons Learned System, Operation UNITED SHIELD

Nonlethal Weapons Joint Mission Area Analysis/Joint Mission Needs Analysis, Dec. 2000

Interservice Nonlethal Individual Weapons Instructor Course (INIWIC) Program of Instruction, Oct.2000

Presidential Decision Directive PDD-39 U.S. Policy on Counterterrorism 21 Jun.1995

PDD-62 Combating Terrorism—Fact Sheet, 22 May 1998

The Federal Response Plan Notice of Change 7, Feb.1997; FEMA, Number 229 Change 11

Federal Response Plan, Apr 1999

GLOSSARY

PART I – ABBREVIATIONS AND ACRONYMS

A

ACOG	advanced combat optical gunsight
ADS	aerial delivery system
AFDC	Air Force Doctrine Center
AFI	Air Force instruction
AFMAN	Air Force manual
ALSA	Air Land Sea Application
AMC	Army Materiel Command
AO	area of operation
AR	Army regulation
ATSC	Army Training Support Center

B

bldg	building
BLT	battalion landing team
bn	battalion

C

C4ISR	command, control, communications, computers, intelligence, surveillance, and reconnaissance
CA	civil affairs
cal	caliber
CANS	compressed air net system
CDC	crowd disposal cartridge
CMC	Commandant of the Marine Corps
COB	close of business
COC	combat operations center
COMDTPUB	Commandant, Coast Guard Publication
COMDTINST	Commandant, Coast Guard Instruction
CONUS	continental United States
COTS	commercial off-the-shelf
CP	command post
CQB	close quarters battle
CS	O-Chlorobenzyl-Malononitrile (riot control agent (tear gas))
CSSOC	combat service support operations center
cu ft	cubic feet

D

DA	Department of the Army
dB	decibel
DECON	decontamination
demo	demonstration
DM	designated marksman
DOD	Department of Defense
DODD	Department of Defense Directive
DODI	Department of Defense Instruction
DODIC	Department of Defense Identification Code
DODAC	Department of Defense Ammunition Code
DOT	Department of Transportation
DP	downed person
DPB	deployable pursuit boat
DTD	dated

E

EADBH	Eagle Duty Baton Holder
EEZ	exclusive economic zone
EOD	explosive ordnance disposal

F

FAST	Fleet Antiterrorism Security Team
FDP	food distribution point
FLEX	forearm/ankle cuff
FM	field manual
four M's	mind, mass, movement, muscle
FSE	fire support element
ft	feet

G

G1	Army or Marine Corps component manpower or personnel staff officer (Army division or higher staff, Marine Corps brigade or higher staff)
G2	Army or Marine Corps component intelligence staff officer (Army division or higher staff, Marine Corps brigade or higher staff)
G3	Army or Marine Corps component operations staff officer (Army division or higher staff, Marine Corps brigade or higher staff)

G4	Army or Marine Corps component logistics staff officer (Army division or higher staff, Marine Corps brigade or higher staff)
ga	gauge
GOTS	government off-the-shelf
H	
HMMWV	high mobility multipurpose wheeled vehicle
I	
IADS	Integrated Air Defense System
IAW	in accordance with
ID	identify/identification
IFF	identification, friend or foe
in.	inch
INIWIC	Interservice Nonlethal Individual Weapons Instructor Course
IPB	intelligence preparation of the battlespace
ITRO	Interservice Training Organization
J	
JMAA	Joint Mission Area Analysis
JTTP	joint tactics, techniques, and procedures
K	
KFOR	Kosovo forces
L	
LBE	load-bearing equipment
LOAC	Law of Armed Conflict
LP	listening post
LVOSS	light visual optical screening system
M	
M16	M16A2 rifle
M203	M203 Grenade Launcher
MARCORSYCOM	Marine Corps System Command
MCCDC	Marine Corps Combat Development Command
MCCM-GE	modified crowd control munition—ground emplaced
MCO	Marine Corps Order
MCPDS	Marine Corps Doctrinal Publication System
MCRP	Marine Corps reference publication

MCSF	Marine Corps security force
MCWP	Marine Corps warfighting publication
MDST	Monadnock Defensive Tactics System
MEDEVAC	medical evacuation
MEF	Marine expeditionary force
METT-T	mission, enemy, terrain and weather, troops and support available – time available
METT-TC	mission, enemy, terrain and weather, troops and support available – time available and civil considerations
MEU	Marine expeditionary unit
MIO	Maritime interception operation
MLEM	Maritime Law Enforcement Manual
MILSTRIP	military
mm	millimeter
MNS	mission need statement
MOAB	management of aggressive behavior
MOOTW	military operations other than war
MOS	military occupational specialty
MOUT	military operations in urban terrain
MP	military police
mph	miles per hour
MRE	meal, ready to eat
MTTP	multi-Service tactics, techniques, and procedures

N

NATO	North Atlantic Treaty Organization
NAVSOP	Navy standing operating procedures
NDC	Naval Doctrine Command
NLFC	nonlethal force capabilities
NLW	nonlethal weapon
NSN	national stock number
NSWC	Naval Surface Warfare Command
NWDC	Naval Warfare Development Command
NWP	Naval Warfare Publication

O

OC	Oleoresin Capsicum
OCAT	Oleoresin Capsicum aerosol training
OCONUS	outside the continental United States
OD	olive drab
OP	observation post

OPFOR	opposing force

P

PA	public affairs
PAO	public affairs officer
PASGT	personnel armor system, ground troops
POI	program of instruction
PSYOP	psychological operations
PT	physical training
PVAB	portable vehicle arresting barrier
PVC	polyvinyl chloride

Q

QRF	quick reaction force

R

R&D	research and development
RCA	riot control agent
RGES	running gear entanglement system
ROE	rules of engagement
RRF	rapid reaction force

S

S1	battalion or brigade manpower or personnel staff officer (Army; Marine Corps battalion or regiment)
S2	battalion or brigade intelligence staff officer (Army; Marine Corps battalion or regiment)
S3	battalion or brigade operations staff officer (Army; Marine Corps battalion or regiment)
S4	battalion or brigade logistics staff officer (Army; Marine Corps battalion or regiment)
SASO	stability and support operations
SB	shield bearer
sec	second
SECNAVINST	Secretary of the Navy instruction
SG	shotgun
SJA	staff judge advocate
SNCO	staff noncommissioned officer
SOP	standard operating procedure
SOTG	Special Operations Training Group
SROE	standing rules of engagement
SSC	small-scale contingency

T

TC	training circular
TF	task force
T/O	table of organization
TOC	tactical operations center
TRADOC	United States Army Training and Doctrine Command
TTP	tactics, techniques, and procedures

U

UN	United Nations
US	United States
USA	United States Army
USAF	United States Air Force
USAMPS	United States Army Military Police School
USCG	United States Coast Guard
USMC	United States Marine Corps
USN	United States Navy
USS	United States Ship

V

VCR	videocassette recorder
VLNL	vehicle launched nonlethal grenade
VMNLGL	vehicle mounted nonlethal grenade launcher

W

w/	with
WMD	weapons of mass destruction

PART II – TERMS AND DEFINITIONS

Air Force specialty code (AFSC). Primary job of a U.S. Air Force officer or enlisted person. Analogous to the Army and Marine Corps military occupational skill (MOS).

area of operation (AO). An operational area defined by the joint force commander for land and naval forces. Areas of operation do not typically encompass the entire operational area of the joint force commander, but should be large enough for component commanders to accomplish their missions and protect their forces. (JP 1-02)

battalion landing team (BLT). In an amphibious operation, an infantry battalion normally reinforced by necessary combat and service elements; the basic unit for planning an assault landing. (JP 1-02)

civil affairs. Designate active and Reserve Component forces and units organized, trained, and equipped specifically to conduct civil affairs activities and to support civil-military operations. (JP 1-02)

civil-military operations. The activities of a commander that establish, maintain, influence, or exploit relations between military forces, governmental and nongovernmental civilian organizations and authorities, and the civilian populace in a friendly, neutral, or hostile operational area in order to facilitate military operations, to consolidate and achieve operational U.S. objectives. Civil-military operations may include performance by military forces of activities and functions normally the responsibility of the local, regional, or national government. These activities may occur prior to, during, or subsequent to other military actions. They may also occur, if directed, in the absence of other military operations. Civil-military operations may be performed by designated civil affairs, by other military forces, or by a combination of civil affairs and other forces. (JP 1-02)

combat service support. The essential capabilities, functions, activities, and tasks necessary to sustain all elements of operating forces in theater at all levels of war. Within the national and theater logistics systems, it includes, but is not limited to, that support rendered by service forces in ensuring the aspects of supply, maintenance, transportation, health services, and other services required by aviation and ground combat troops to permit those units to accomplish their missions in combat. Combat service support encompasses those activities at all levels of war that produce sustainment to all operating forces on the battlefield. (JP 1-02)

command and control. The exercise of authority and direction by a properly designated commander over assigned and attached forces in the accomplishment of the mission. Command and control functions are performed through an arrangement of personnel, equipment, communications, facilities, and procedures employed by a commander in planning, directing, coordinating, and controlling forces and operations in the accomplishment of the mission. (JP 1-02)

commercial off the shelf (COTS). Describes items procured for military use through civilian contractors.

core capability. The set of functions that define a significant, stand-alone, operationally effective and suitable military capability such that, should no further development occur, the user will have received a significant capability.

designated marksman (DM). A rifleman stationed in an overwatch position above friendly forces, armed with a standard infantry weapon appropriate to his service that is mounted with a high-powered scope and loaded with lethal ammunition. The designated marksman uses his vantage point to identify agitators to forces below, eliminate lethal threats, and monitor the battlefield.

dispersion. A scattered pastern of hits around the mean point of impact of bombs and projectiles dropped or fired under identical conditions. In chemical and biological operations, the dissemination of agents in liquid or aerosol form. (JP 1-02)

fleet Marine force (FMF). A balanced force of combined arms comprising land, air, and service elements of the U.S. Marine Corps. A fleet Marine force is an integral part of a U.S. Fleet and has the status of a type command. (JP 1-02)

force protection. Security program designed to protect service members, civilian employees, family members, facilities, and equipment, in all locations and situations, accomplished through planned and integrated application of combating terrorism, physical security, operations security, and personal protective services, and supported by intelligence, counterintelligence, and other security programs. (JP 1-02)

government off-the-shelf (GOTS). Analogous to COTS, but describing items already procured by the U.S. government and ready for issue to military units.

host nation. A nation that receives the forces and/or supplies of allied nations, coalition partners, and/or NATO organizations to be located on, to operate in, or to transit through its territory. (JP 1-02)

intelligence preparation of the battlespace (IPB). An analytical methodology employed to reduce uncertainties concerning the enemy, environment, and terrain for all types of operations. Intelligence preparation of the battlespace builds an extensive database for each potential area in which a unit may be required to operate. The database is then analyzed in detail to determine the impact of the enemy, environment, and terrain on operations, and present it in graphic form. Intelligence preparation of the battlespace is a continuing process. Also called IPB. (JP 1-02)

interrogator-translator team (ITT). Marine Corps unit, usually working as a small detachment to a larger infantry unit, highly trained in interrogation techniques and fluent in adversary native language.

joint force commander. A general term applied to a combatant commander, subunified commander, or joint task force commander authorized to exercise combatant command (command authority) or operational control over a joint force. (JP 1-02)

logistics. The science of planning and carrying out the movement and maintenance of forces. In its most comprehensive sense, those aspects of military operations which deal with:

a. Design and development, acquisition, storage, movement, distribution, maintenance, evacuation, and disposition of materiel.

b. Movement, evacuation, and hospitalization of personnel.

c. Acquisition of construction, maintenance, operation, and disposition of facilities.

d. Acquisition of furnishing of services. (JP 1-02)

low-intensity conflict. Political-military confrontation between contending states or groups below conventional war and above the routine, peaceful competition among states. It frequently involves protracted struggles of competing principles and ideologies. Low intensity conflict ranges from subversion to the use of armed force. It is waged by a combination of means employing political, economic, informational, and military instruments. Low-intensity conflicts are often localized, generally in the Third World, but contain regional and global security implications.

NOTE: Also known as small wars or "brushfire wars," low-intensity conflicts are rebellions, insurrections, and the like characterized by adversaries dedicated to a cause but lacking command and control structure and modern equipment.

Marine expeditionary force (MEF). The largest of the Marine air-ground task force (MAGTF) and the Marine Corps principal warfighting organization, particularly for larger crisis or contingencies. It is task-organized around a permanent command element and normally contains one or more Marine divisions, Marine aircraft wings, and Marine force service support groups. The Marine expeditionary force is capable of missions across the range of military operations, including amphibious assault and sustained operations ashore in any environment. It can operate from a sea base, a land base, or both. Also called MEF. See also aviation combat element; combat service support element; command element; Marine air-ground task force; Marine expeditionary force (forward); Marine expeditionary unit; special purpose Marine air-ground task force; task force. (JP 1-02)

maritime interception operation (MIO). Means by which an exclusion zone at sea is enforced. Involves the visit of merchant ships, by armed boarding parties if necessary, bound to, through, or out of a defined area, to examine each ship's papers and cargo and to search for evidence of contraband. The provisions must be established by a sanctioning body and must be applied to ships of all nationalities.

military operations other than war (MOOTW). Operations that encompass the use of military capabilities across the range of military operations short of war. These military actions can be applied to complement any combination of the other instruments of national power and occur before, during, and after war. (JP 1-02)

military occupational specialty (MOS). Primary job of military officer or enlisted personnel.

military operations in urban terrain (MOUT). All military actions planned and conducted on a topographical complex and its adjacent natural terrain where man-made construction is the dominant feature. It includes combat-in-cities, which is that portion of MOUT involving house-to-house and street-by-street fighting in towns and cities.

military police (MP). Marine Corps and Army police forces designated by specific MOS.

maritime pre-positioning force operation. A rapid deployment and assembly of a Marine expeditionary force in a secure area using a combination of strategic airlift and forward-deployed maritime pre-positioning ships. See also Marine expeditionary force; maritime pre-positioning ships (JP 1-02 and JP 4-01.6)

nonlethal weapons (NLW). Weapons that are explicitly designed and primarily employed so as to incapacitate personnel or materiel, while minimizing fatalities, permanent injury to personnel, and undesired damage to property and the environment.

a. Unlike conventional weapons that destroy their targets through blast, penetration, and fragmentation, nonlethal weapons employ means other than gross physical destruction to prevent the target from functioning.

b. Nonlethal weapons are intended to have one, or both, of the following characteristics:

(1) They have relatively reversible effects on personnel or materiel.

(2) They affect objects differently within their area of influence. (JP 1-02)

nonlethal weapon capability set. A set of systems containing weapon systems, munitions, and protective equipment. The components are explicitly designed and primarily employed to provide additional protection to friendly personnel and to incapacitate hostile personnel or material, while minimizing fatalities or permanent injury to intended targets and minimal damage to property and the environment. They are not required to have zero probability of producing fatalities or permanent injuries but are designed and employed in a manner that significantly reduces those probabilities over the traditional employment of currently fielded military weapon systems, munitions, and equipment.

nonlethal technology (N-LT). Synonymous with NLW, but also encompassing more advanced technology that is not yet available for use by military forces.

Oleoresin Capsicum (OC). Also known as "pepper spray", OC is made completely from organic materials and is FDA-approved for sale over-the-counter in the United States. Causes severe and immediate burning sensation to mucous membranes when sprayed into face, nose, and eyes.

psychological operations (PSYOP). Planned operations to convey selected information and indicators to foreign audiences to influence their emotions, motives, objective reasoning, and ultimately the behavior of foreign governments, organizations, groups, and individuals. The purpose of psychological operations is to induce or reinforce foreign attitudes and behavior favorable to the originator's objectives. (JP 1-02)

rules of engagement (ROE). Directives issued by competent military authority that delineate the circumstances and limitations under which U.S. forces will initiate and/or continue combat engagement with other forces encountered. See also law of war. (JP 1-02)

table of organization (T/O). Total manpower authorized for specific unit, organized by billet and rank. TO& E is a U.S. Army term which encompasses T/O and T/E.

training aids and devices. Any item developed or procured with the primary intent to assist in training and learning.

weapons of mass destruction (WMD). Weapons capable of a high order of destruction and/or of being used in such a manner as to destroy large numbers of people. Weapons of mass destruction can be high explosives or nuclear, biological, chemical, and radiological weapons, but exclude the means of transporting or propelling the weapon where such means is a separable and deliverable part of the weapon. (JP 1-02)

Index

Made in the USA
Monee, IL
05 November 2026

67811381R00148